AF353163

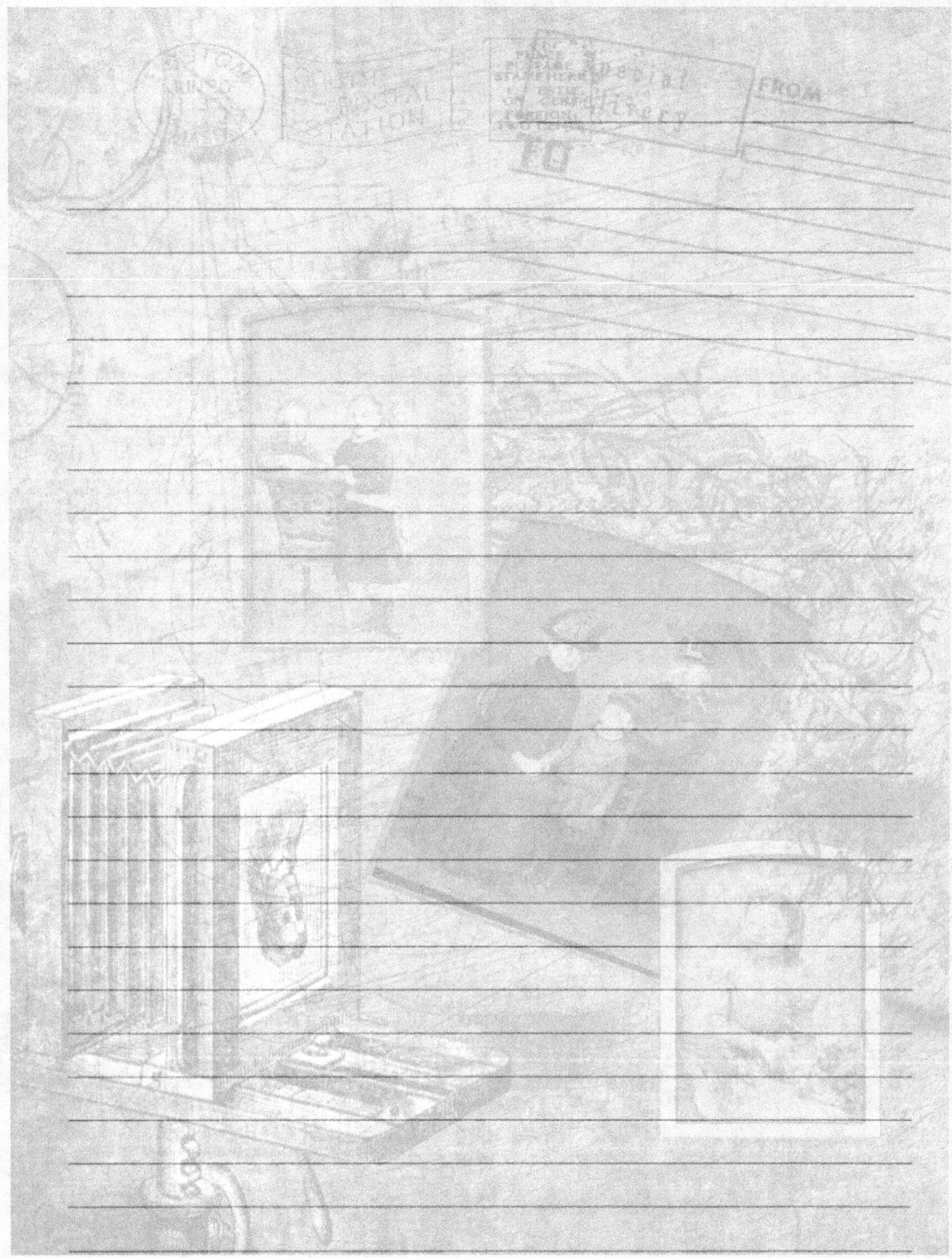

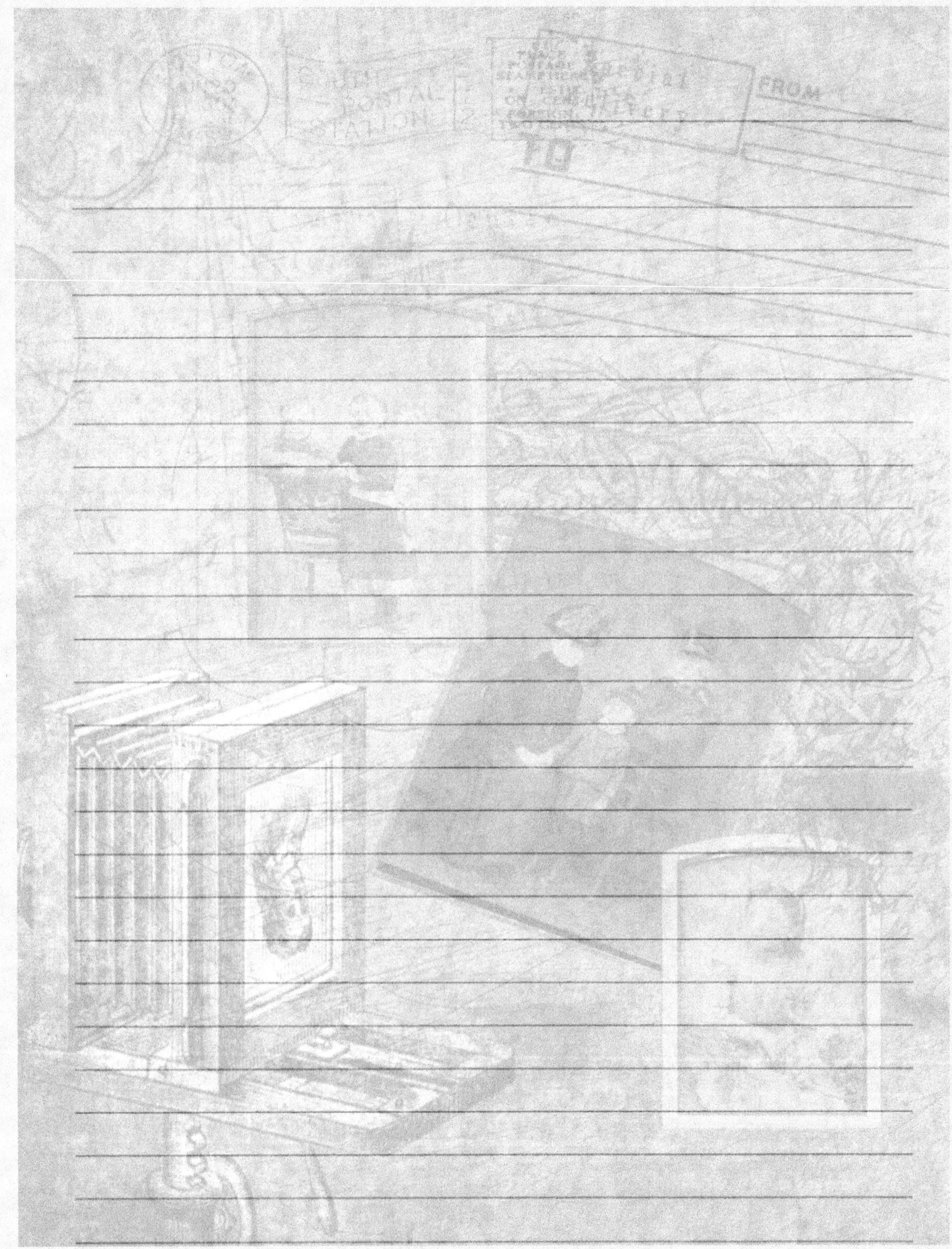

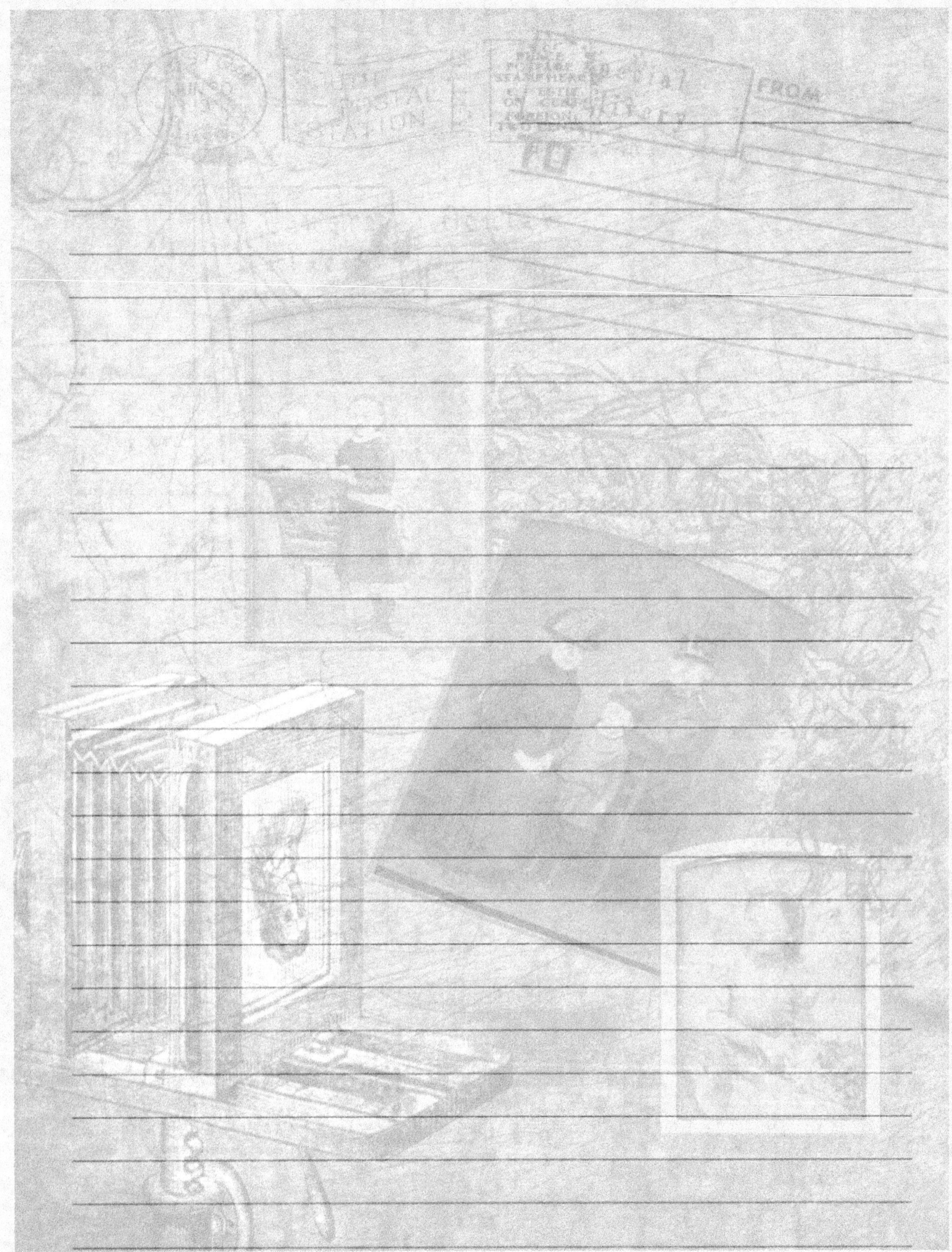

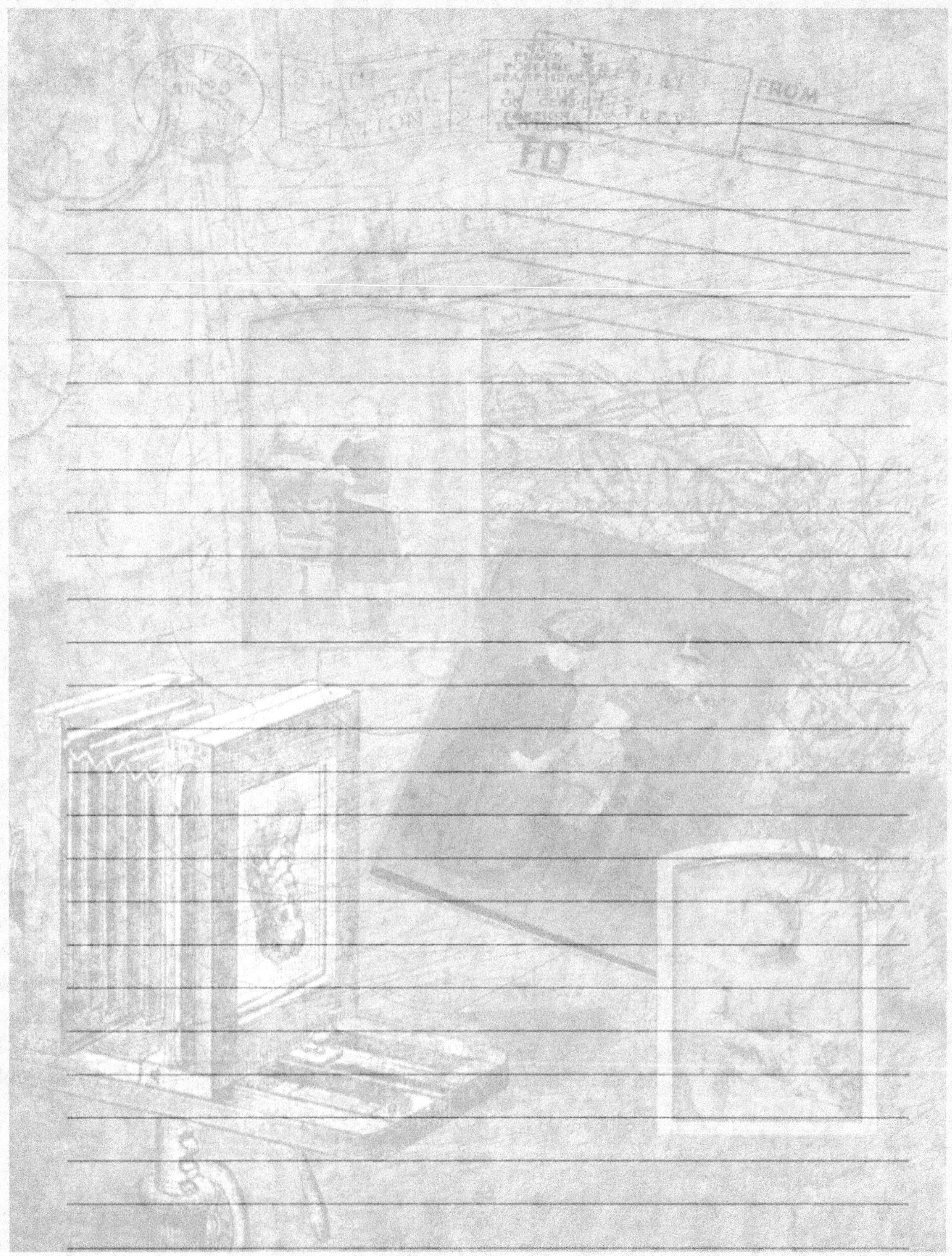

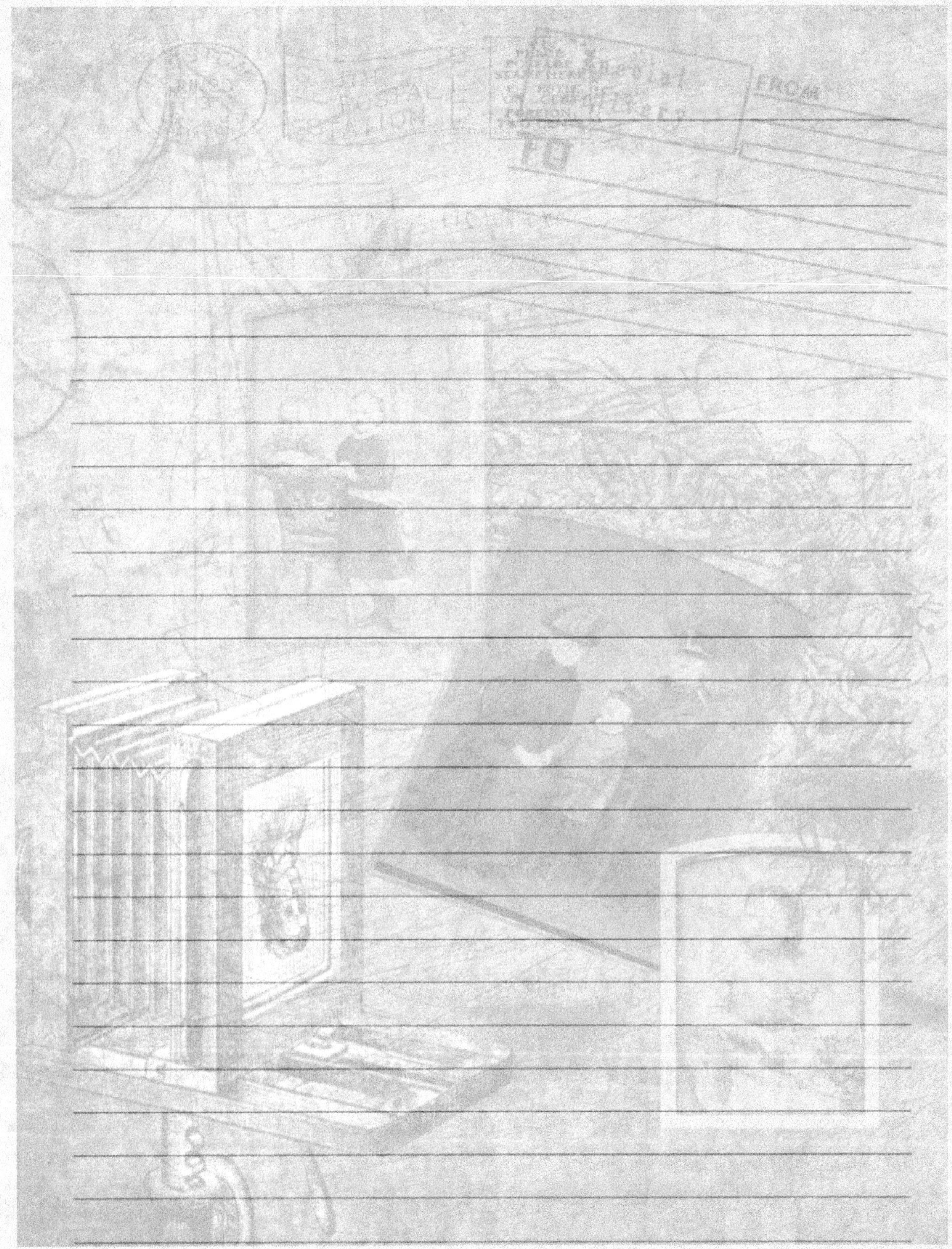

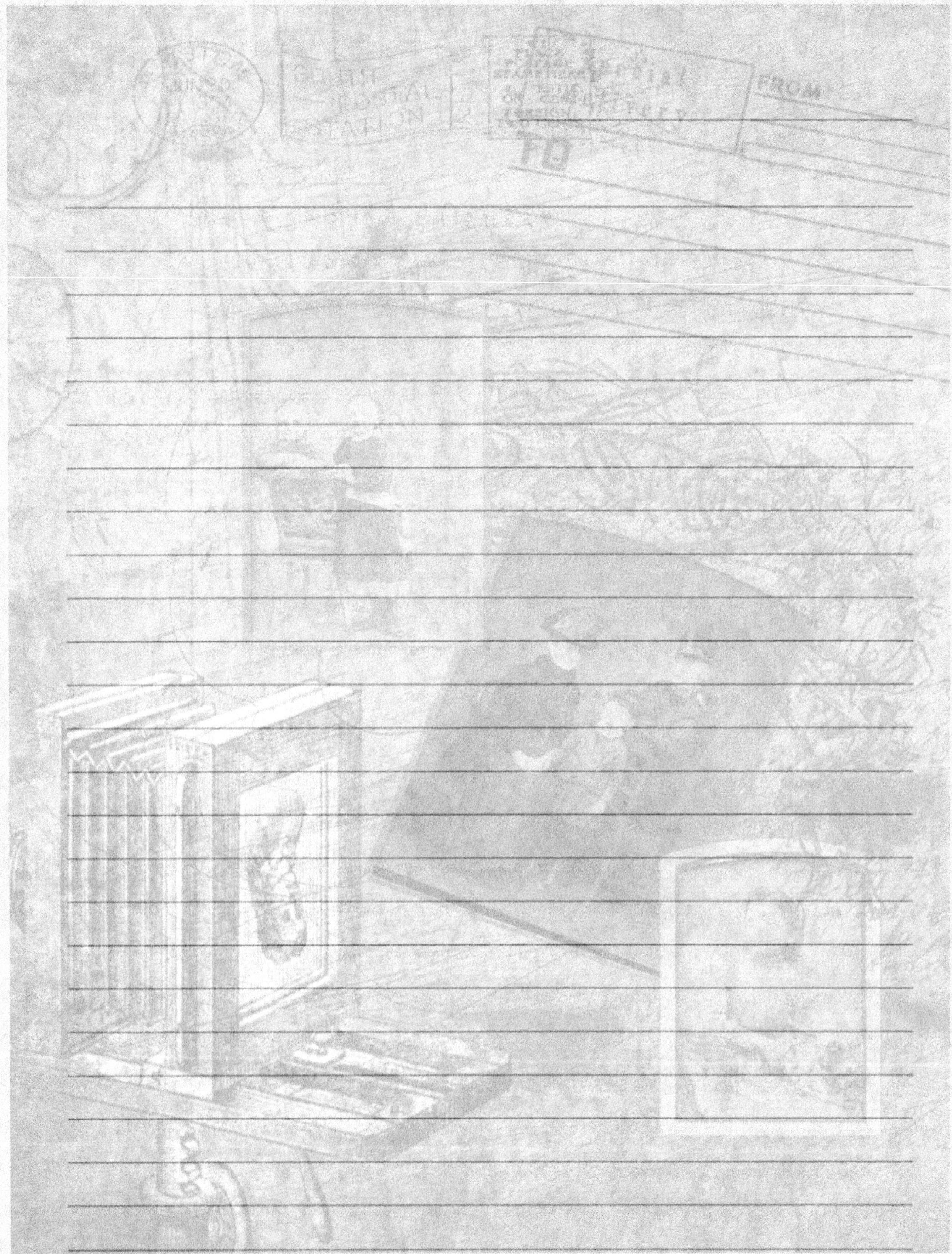

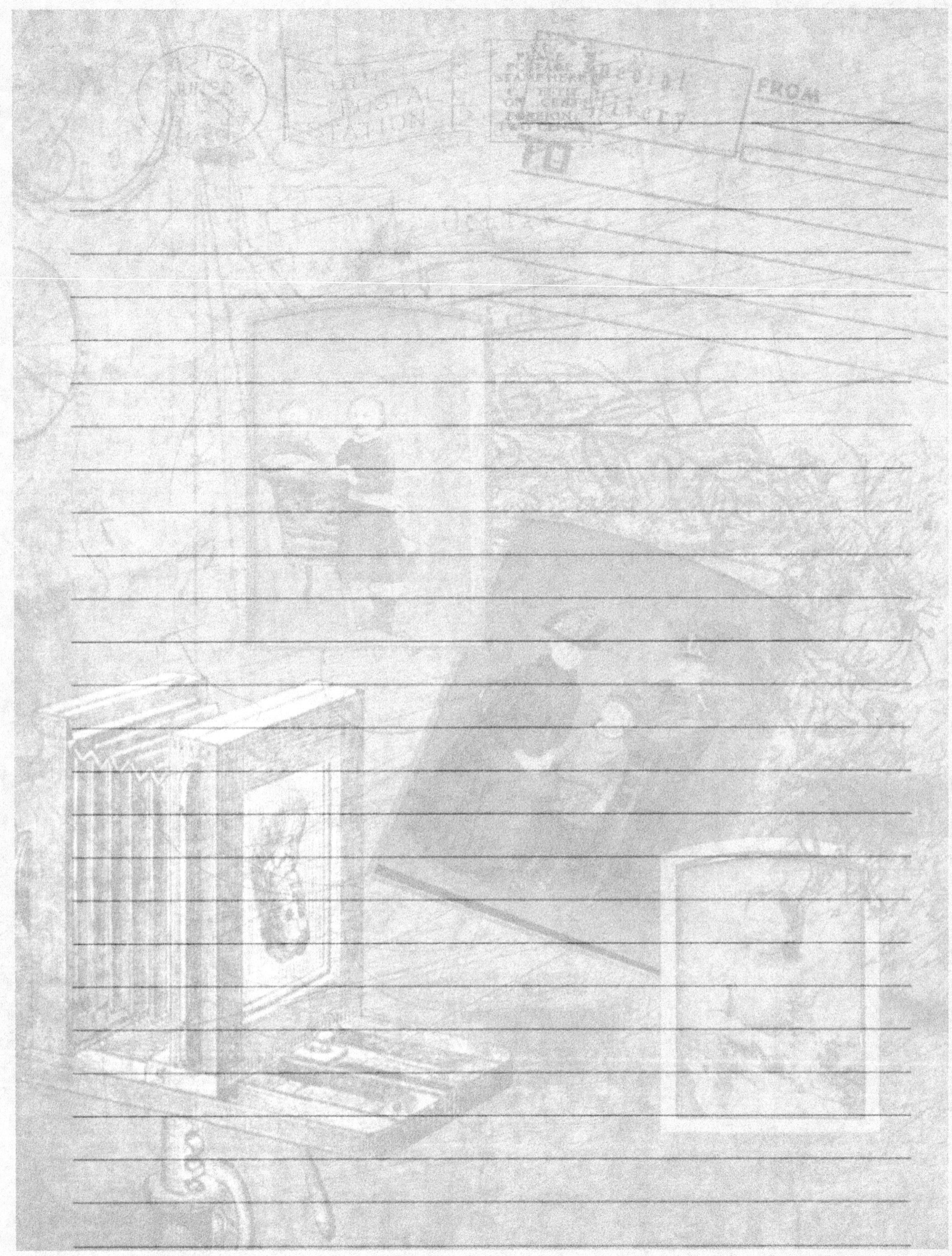

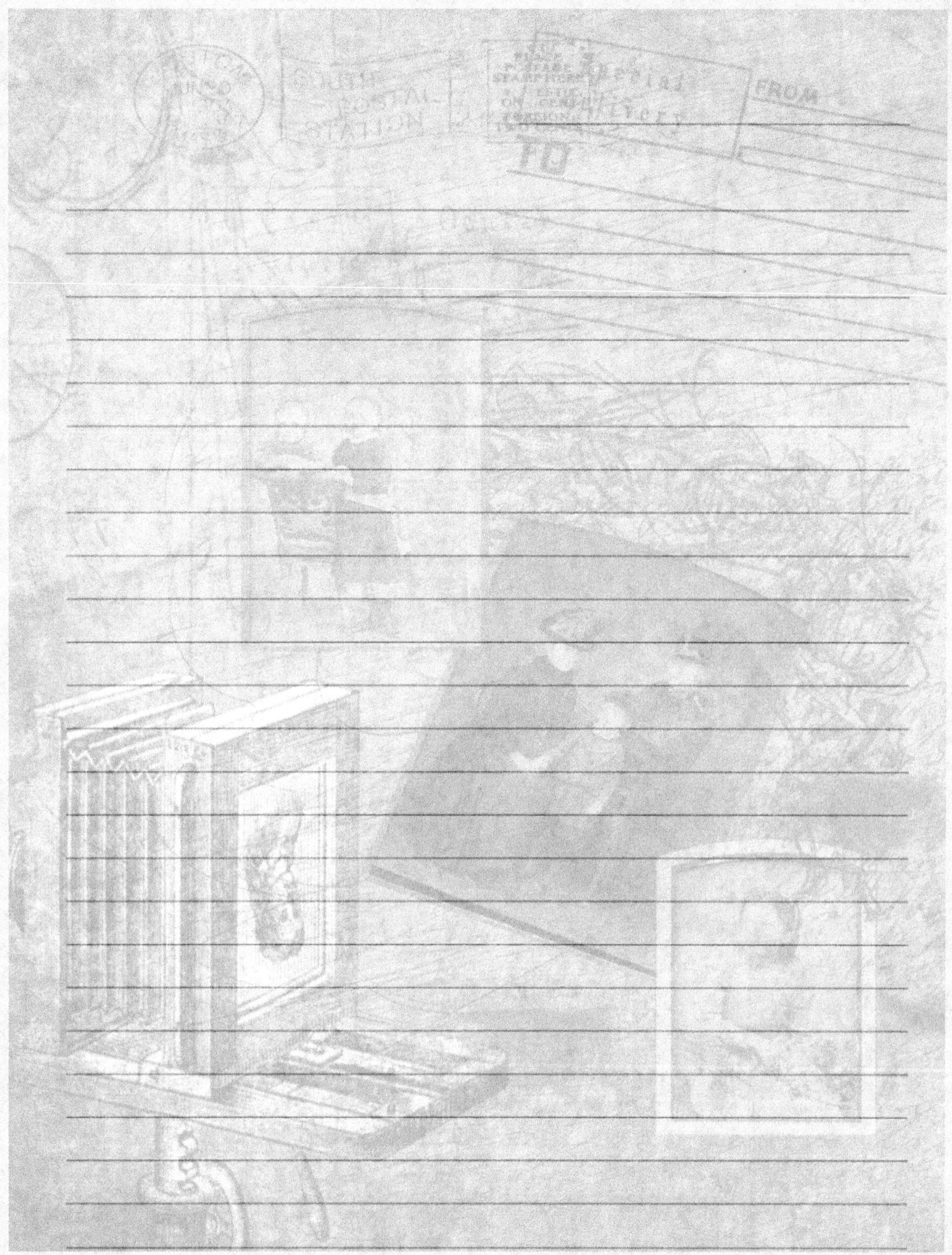

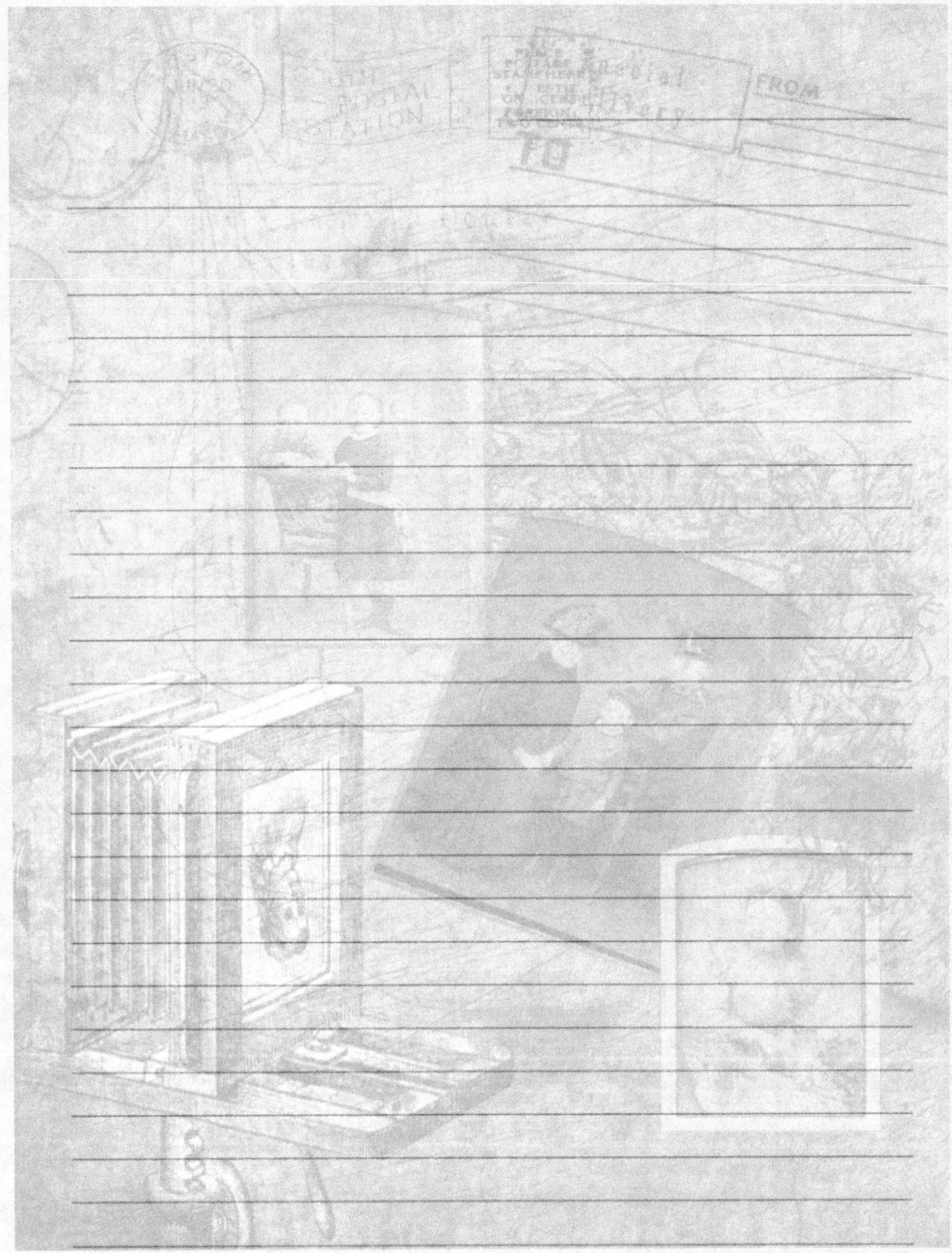

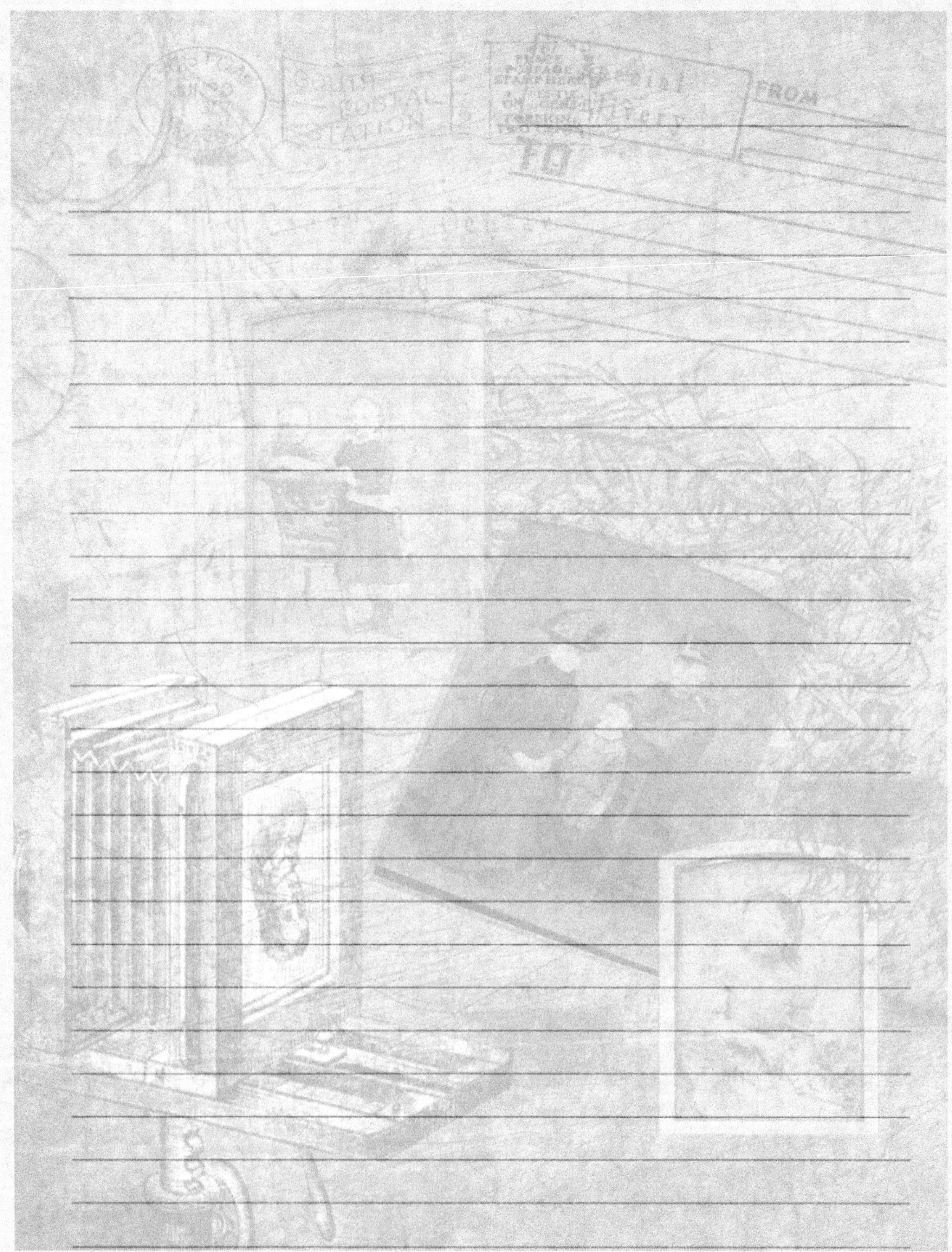

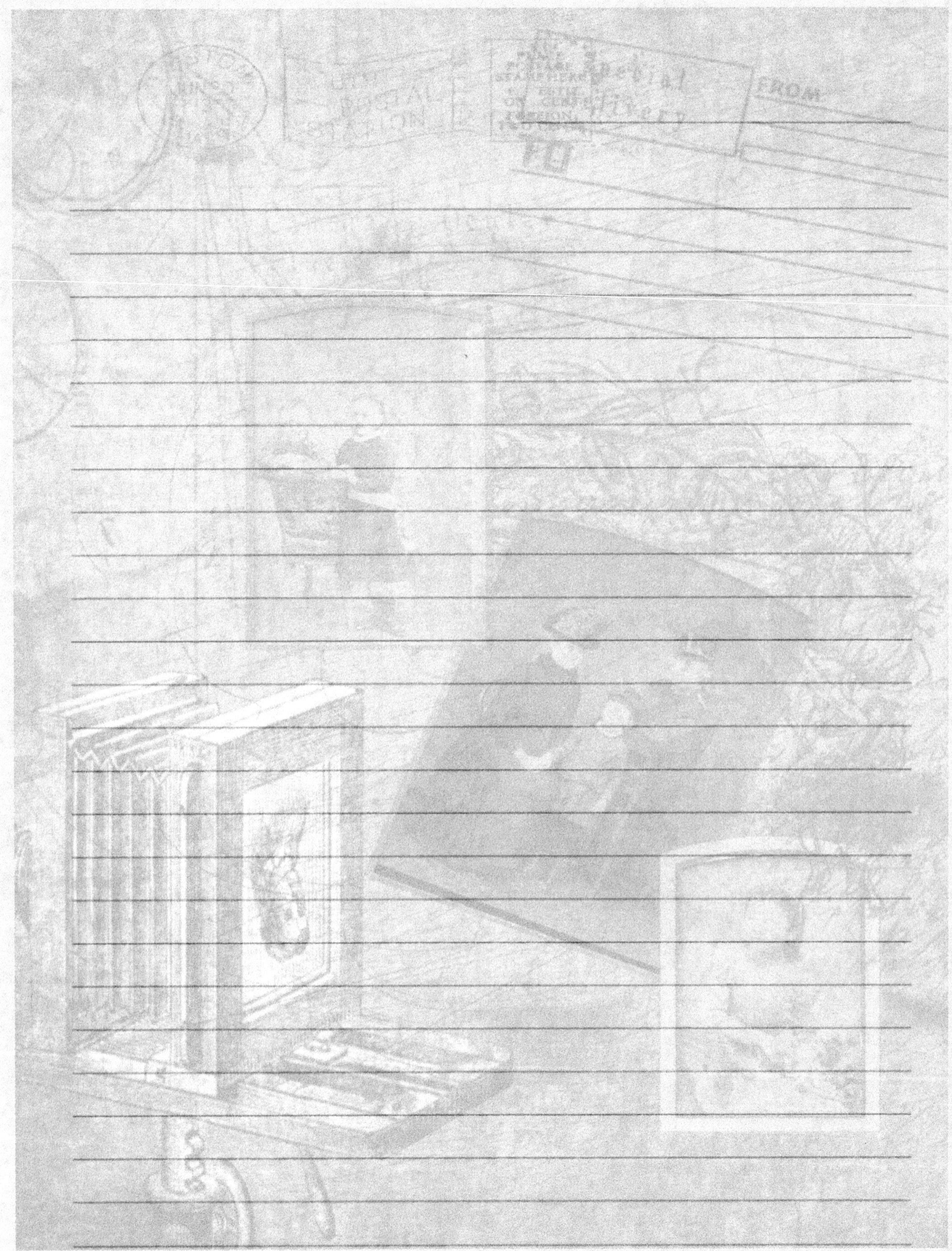

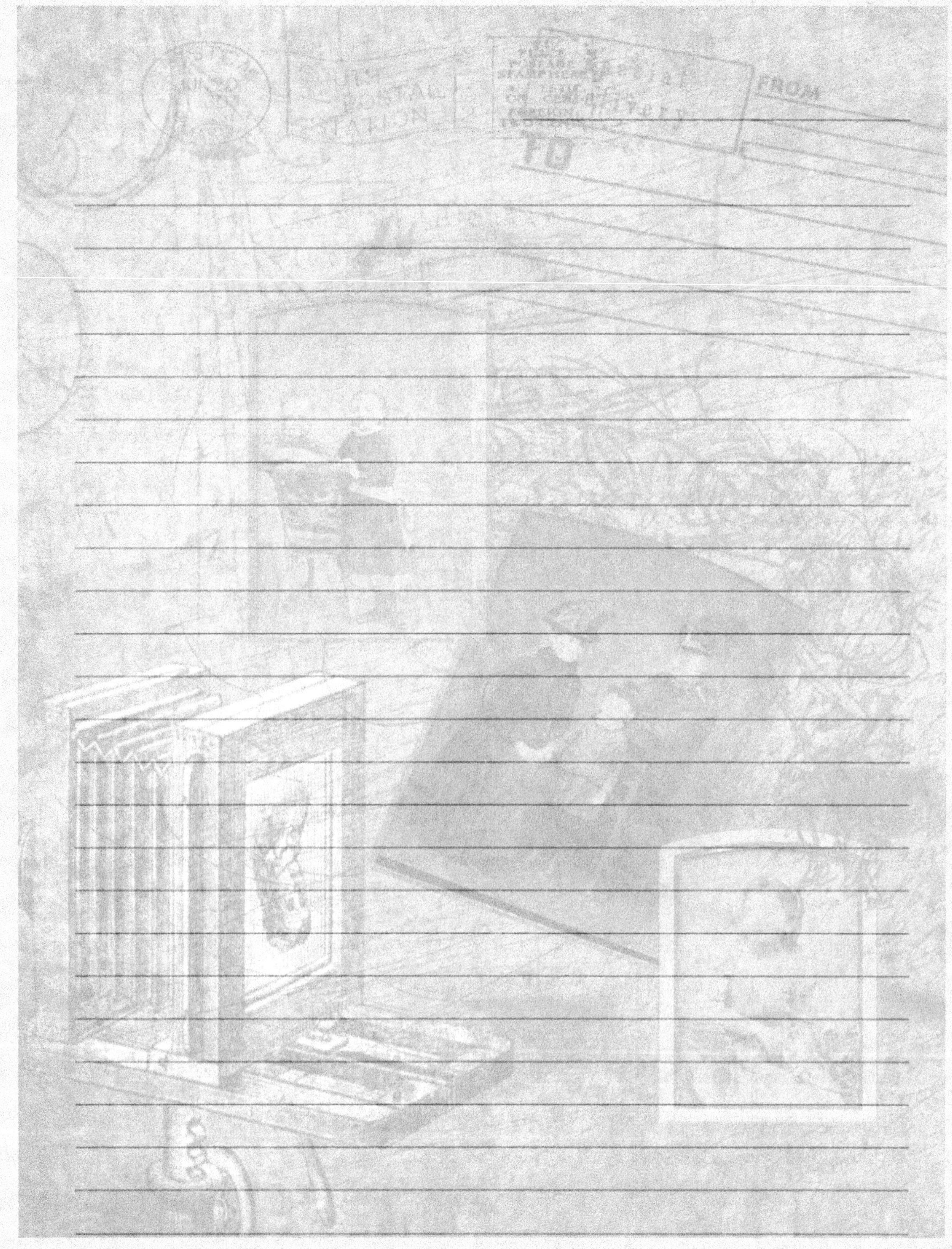

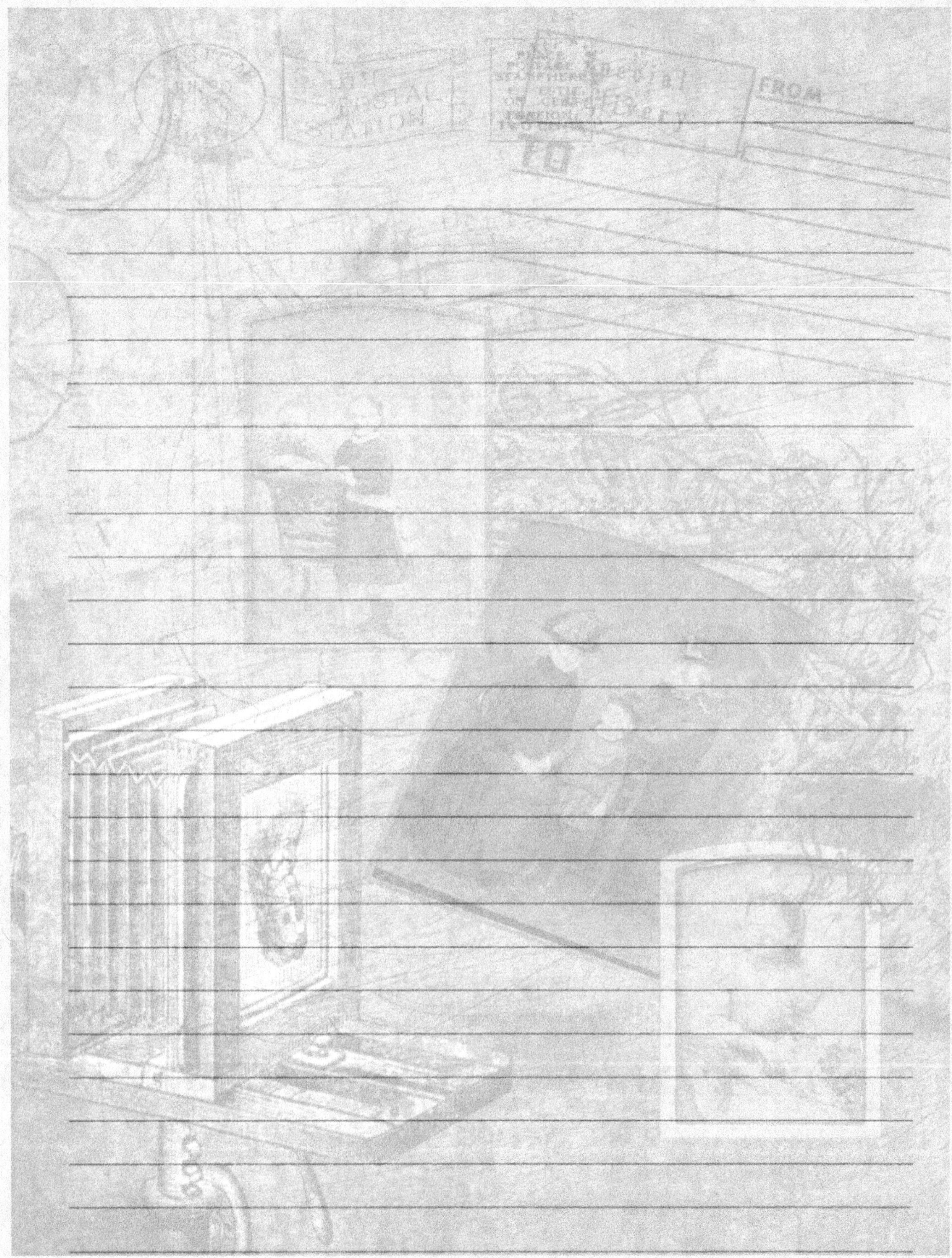

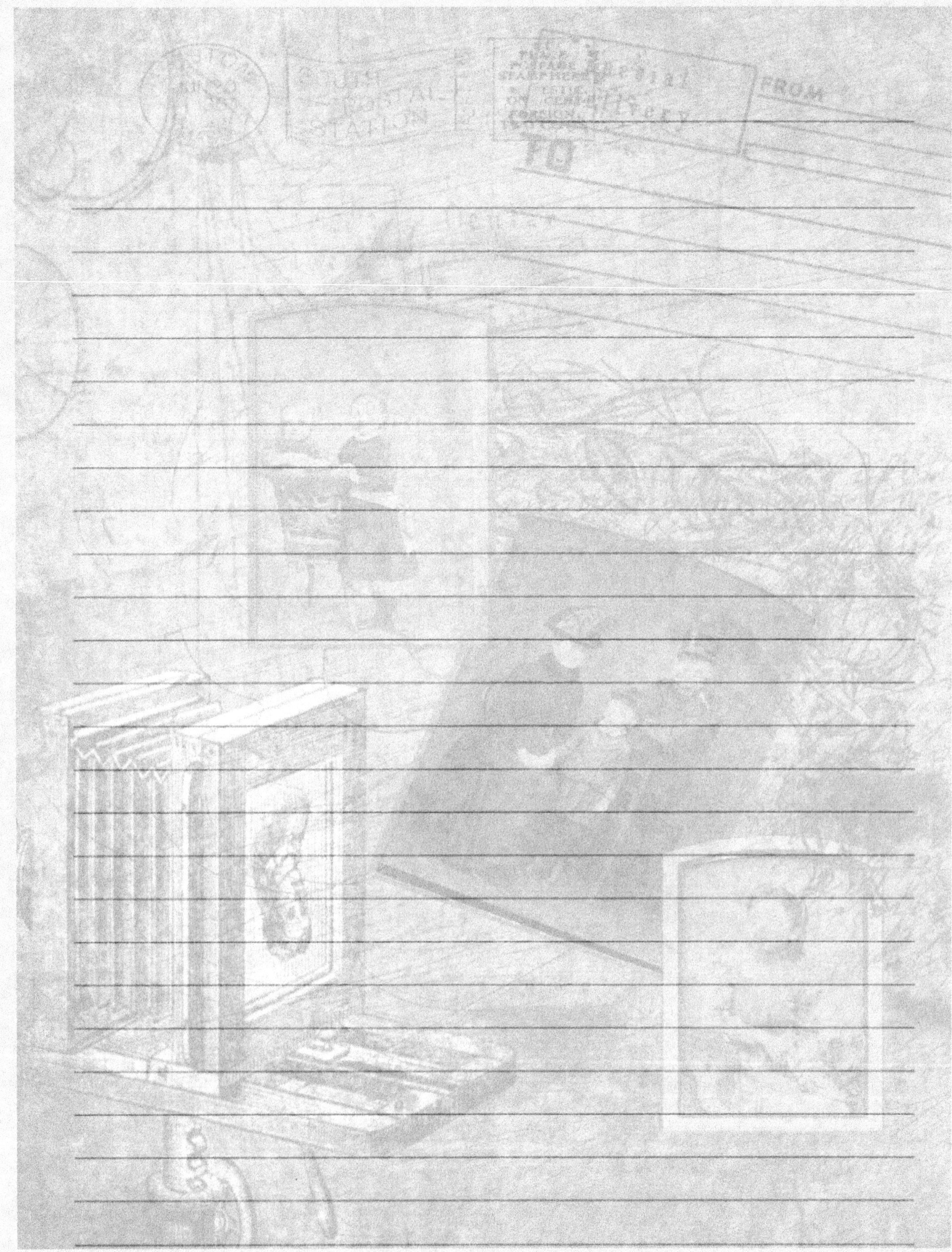

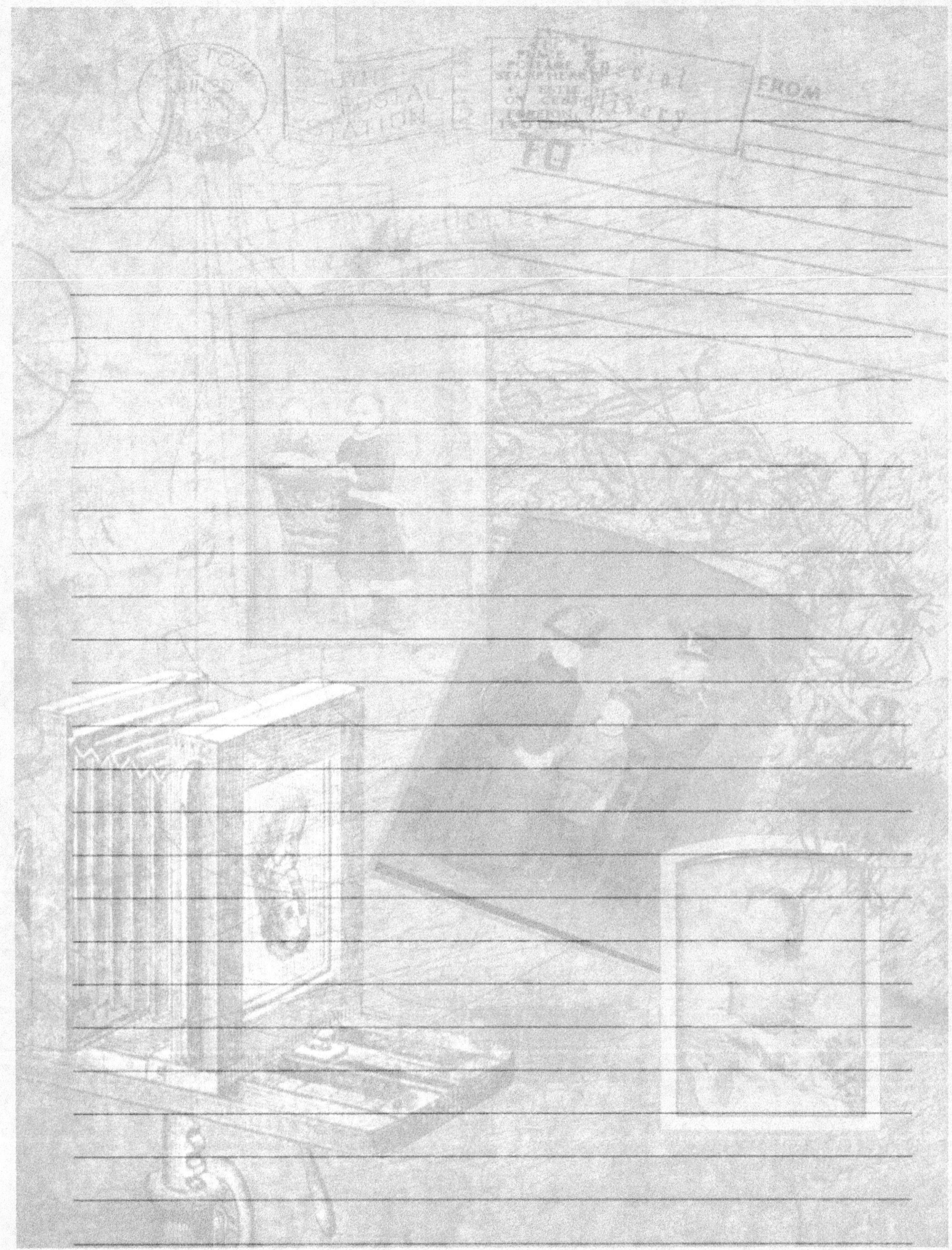

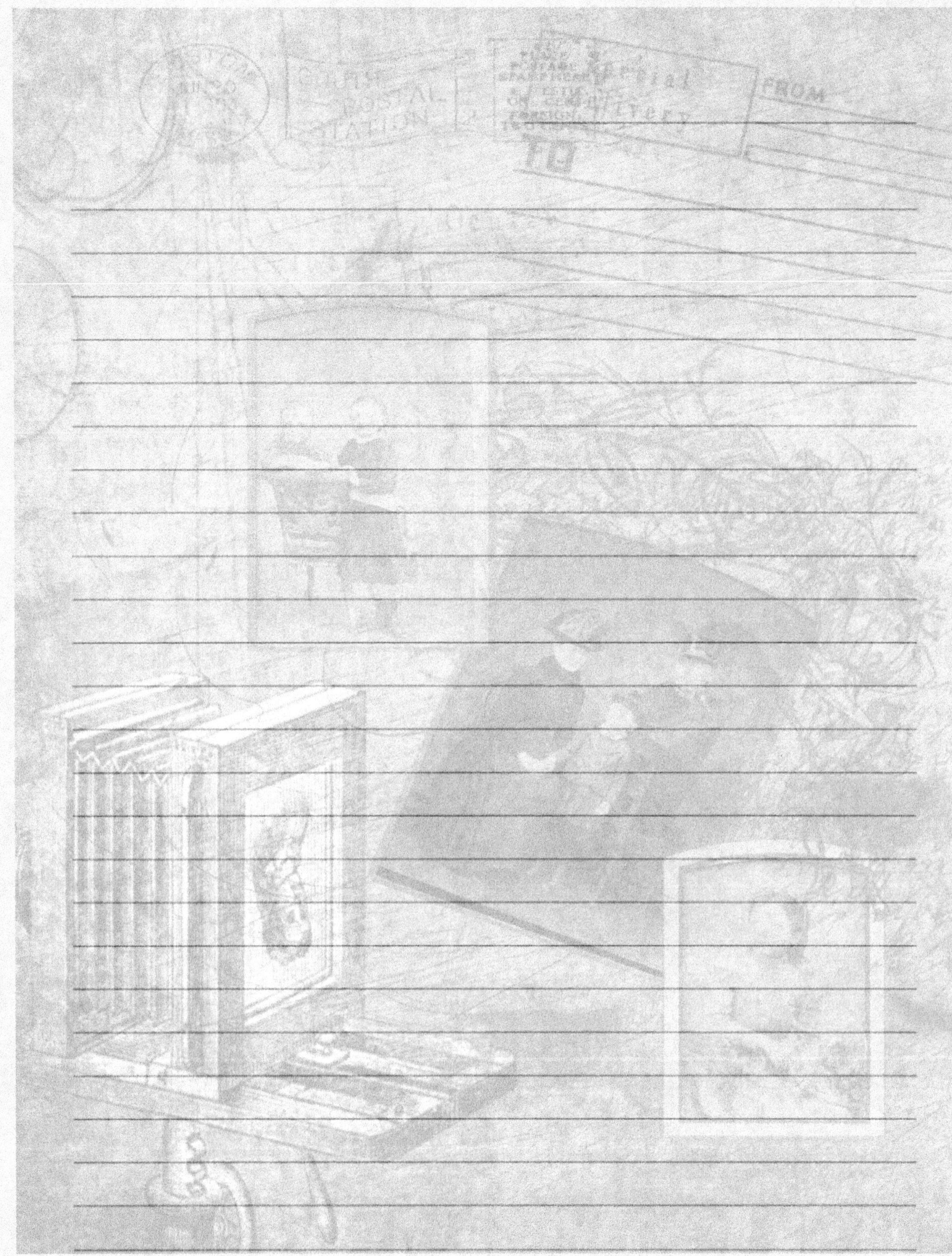

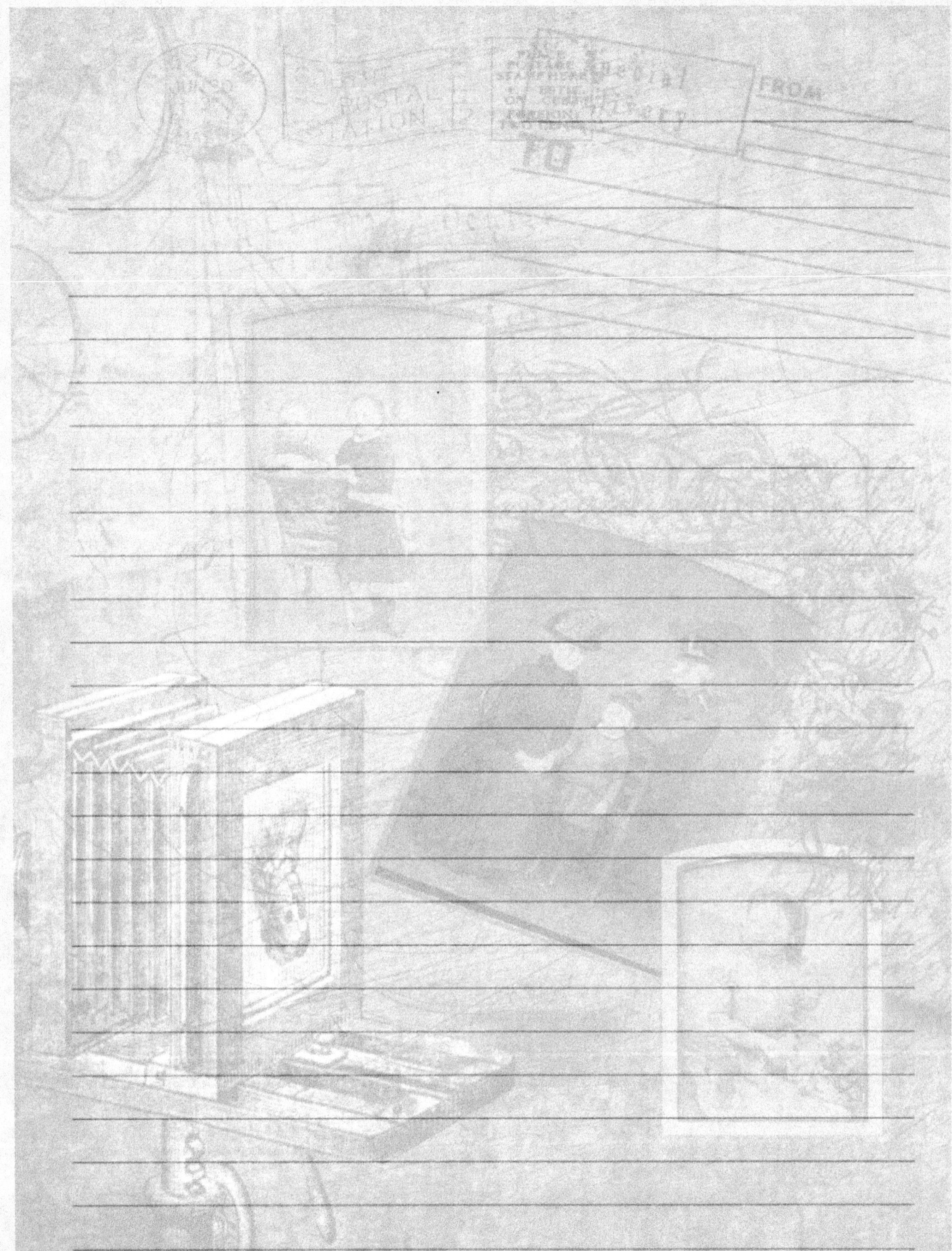

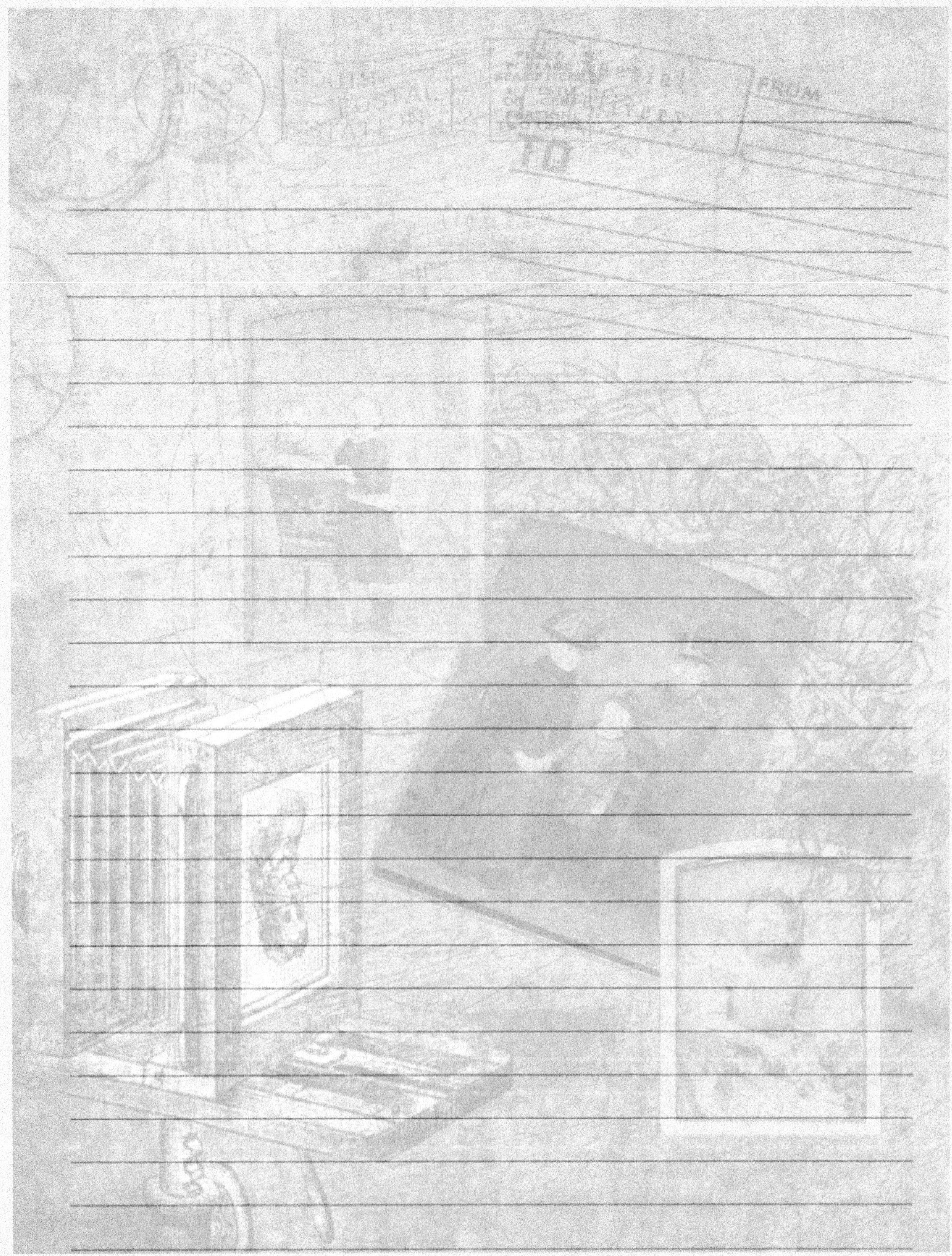

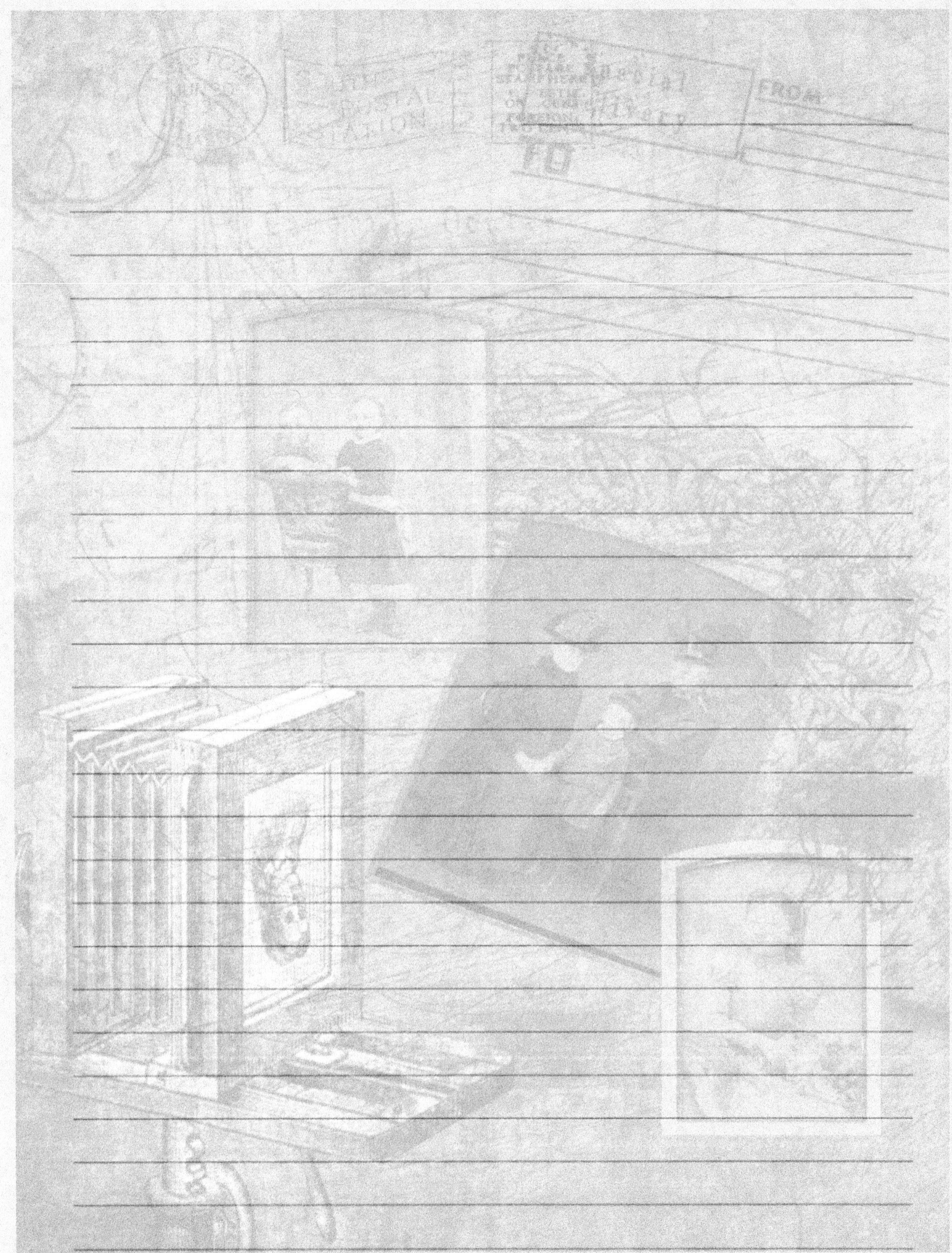

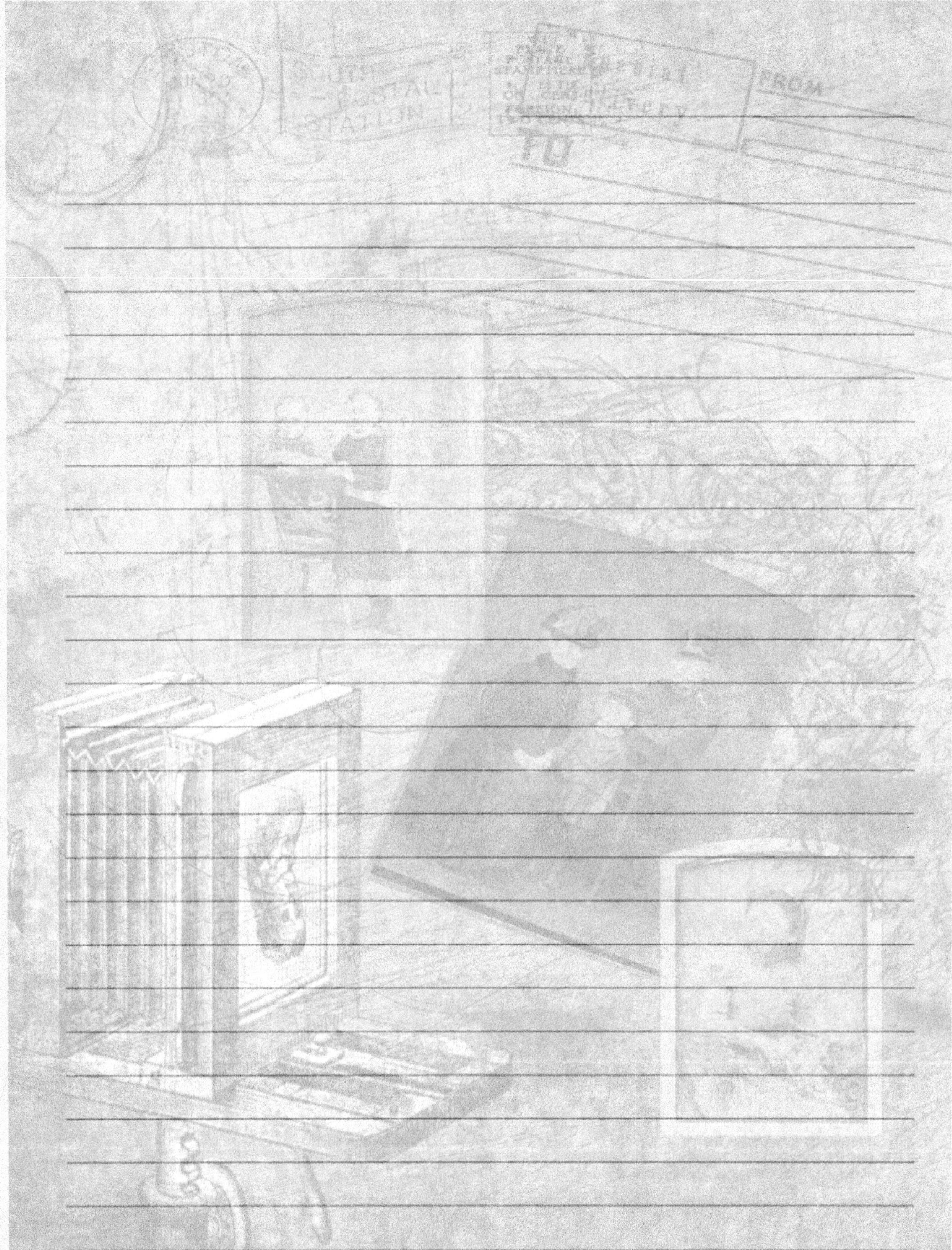

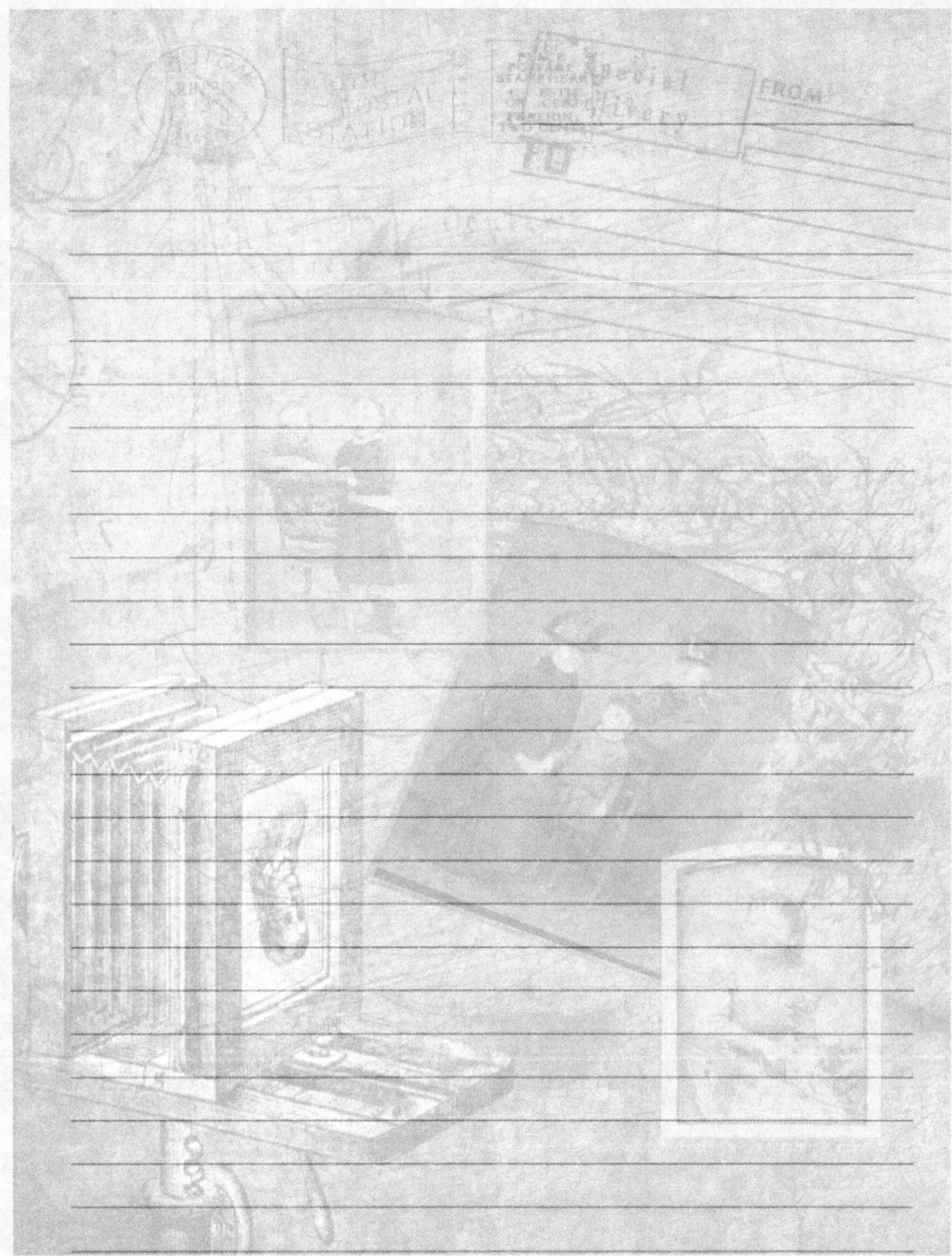

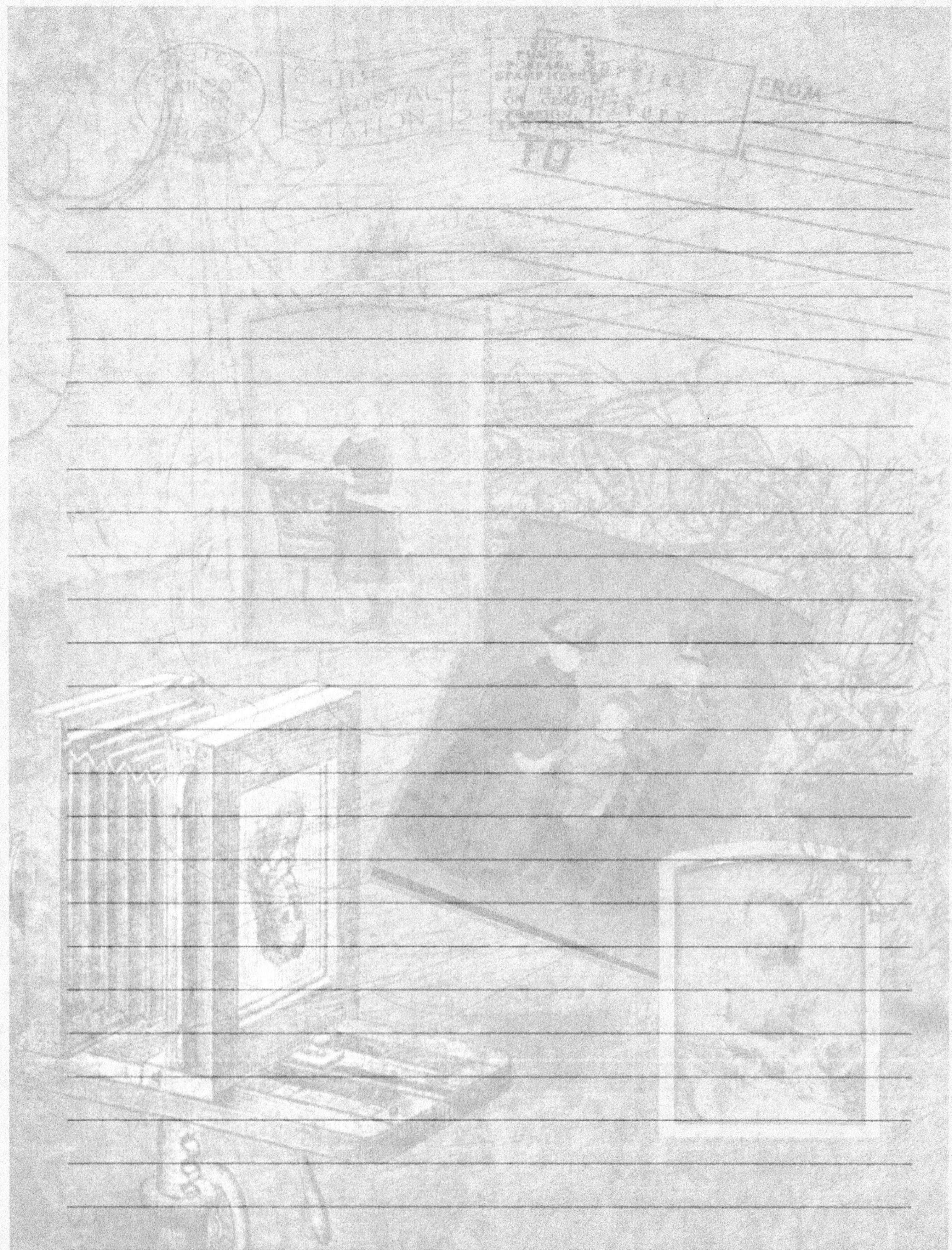

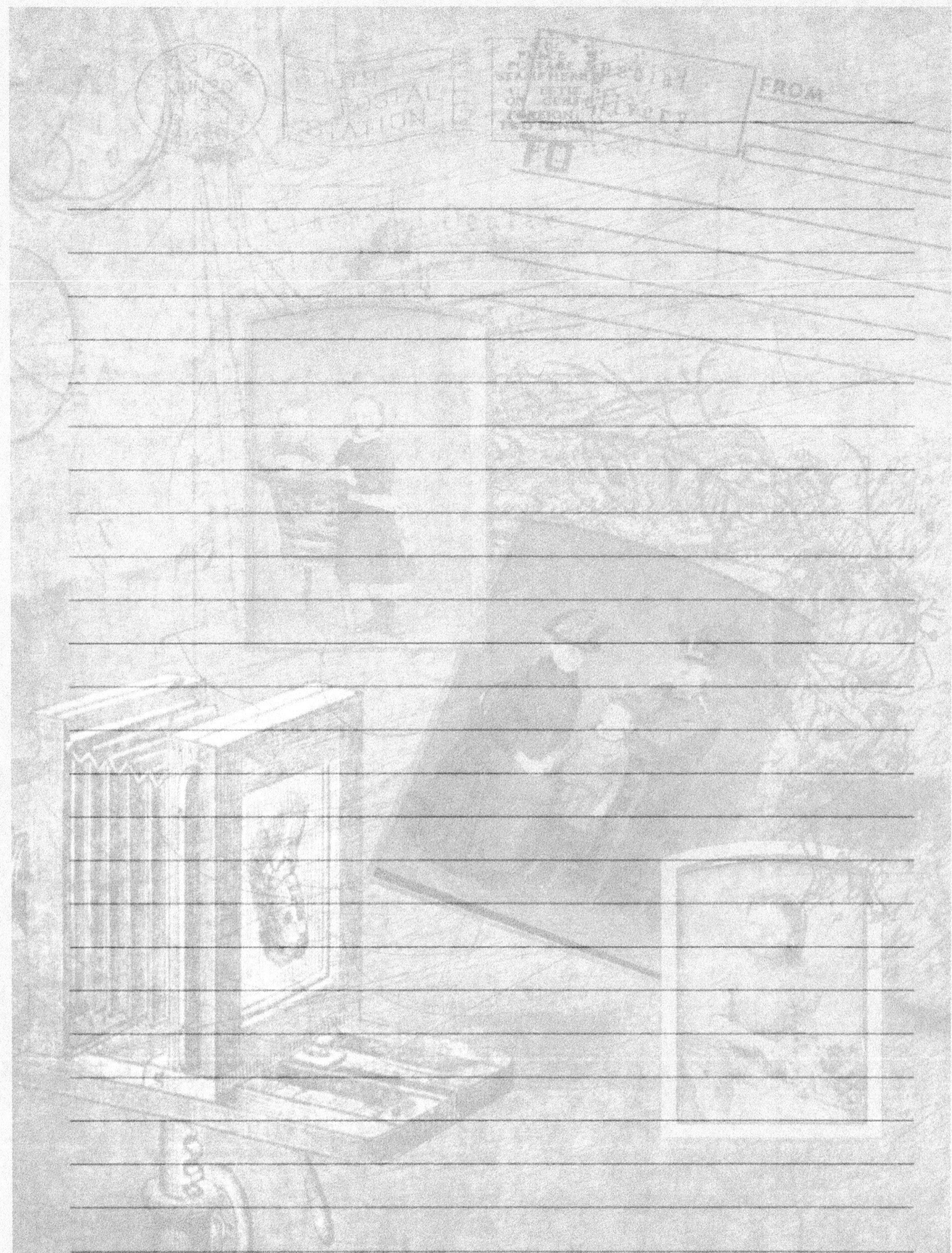

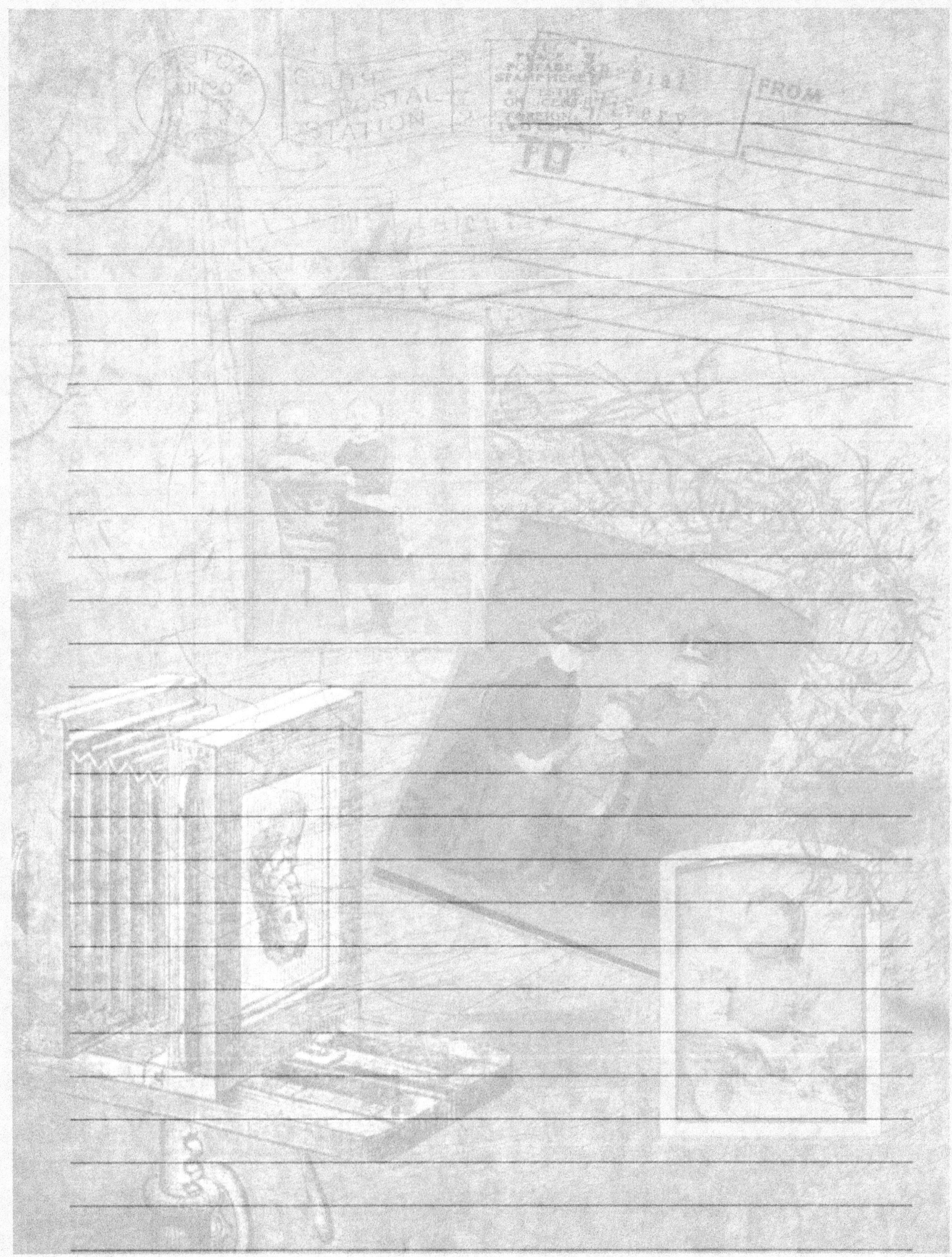

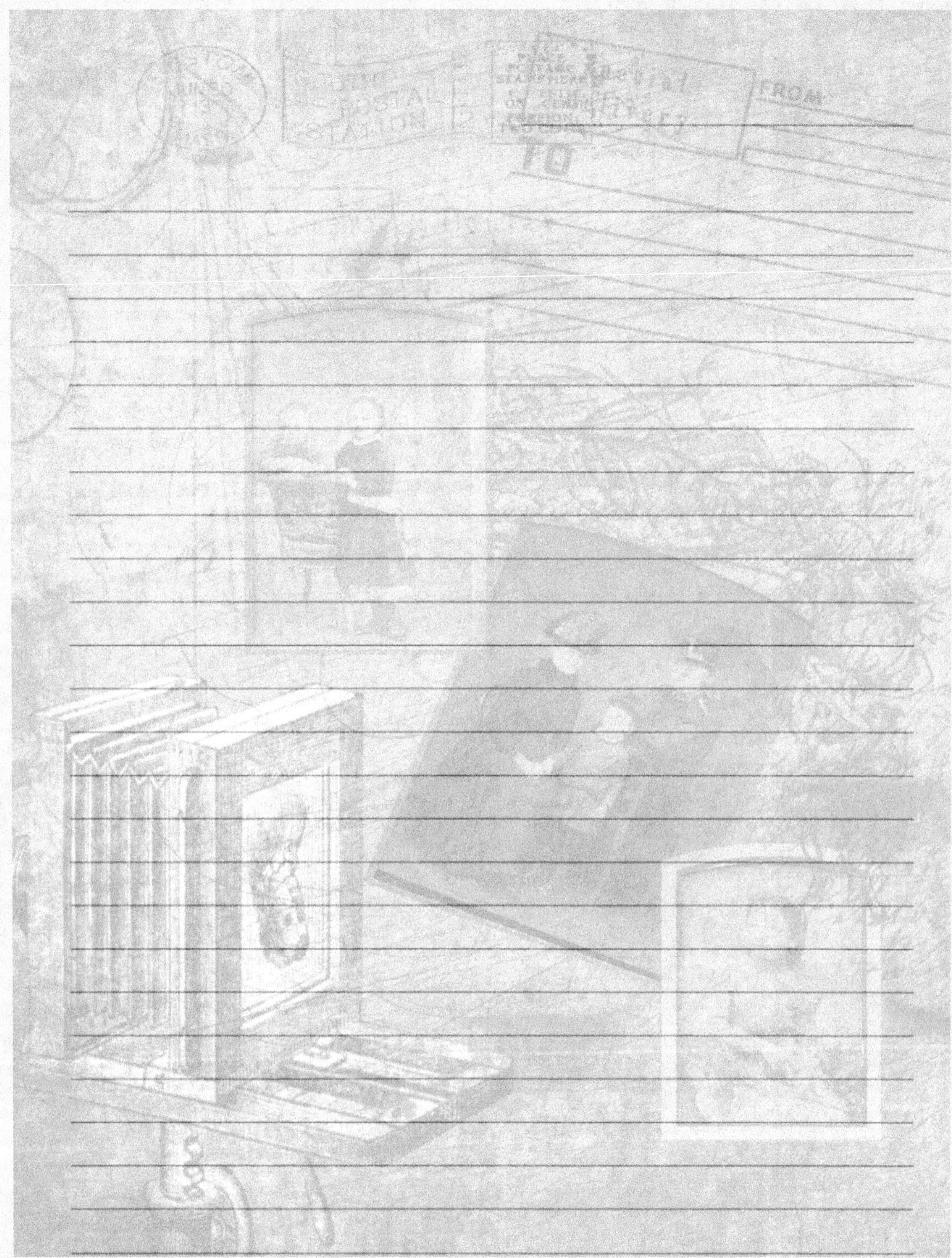

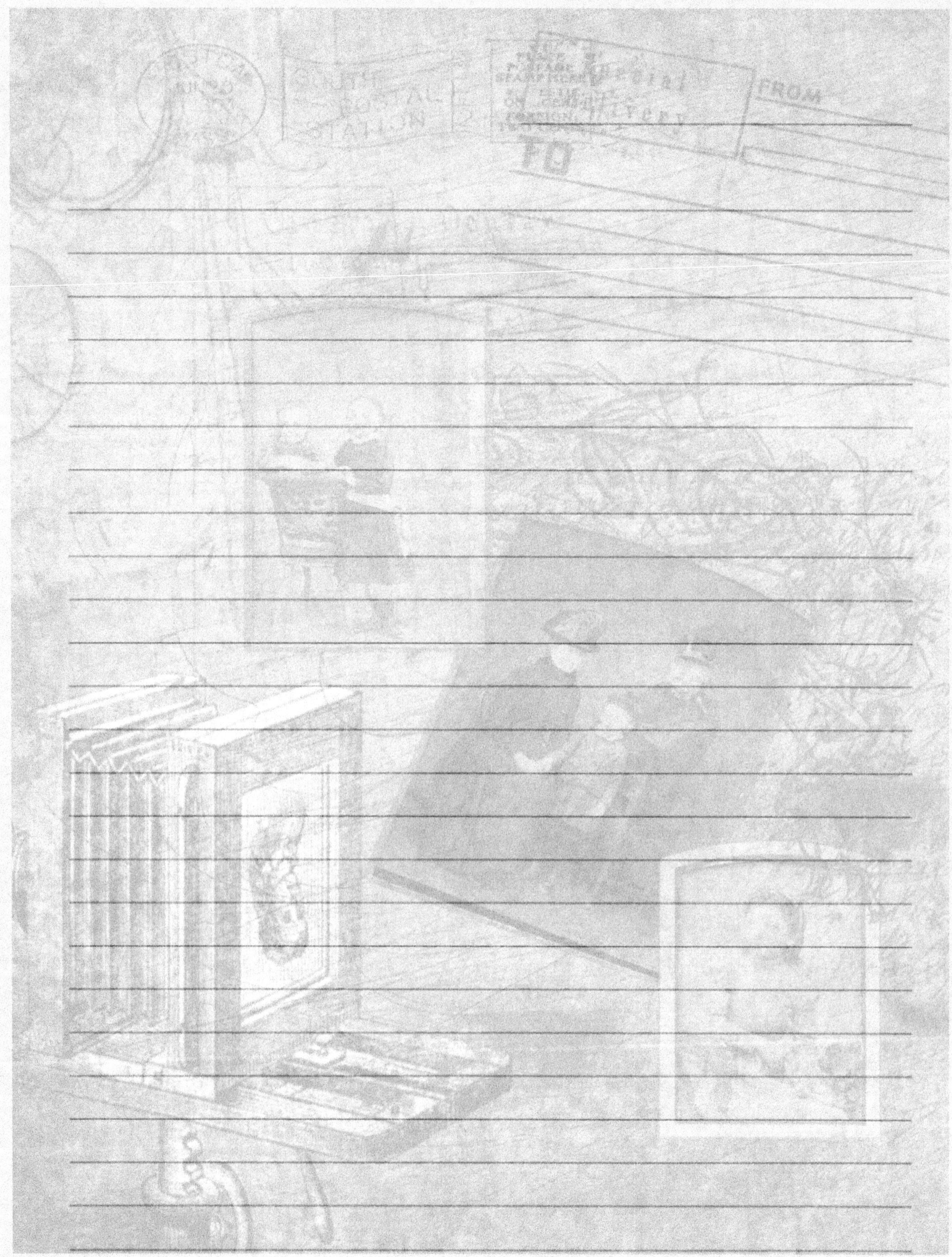

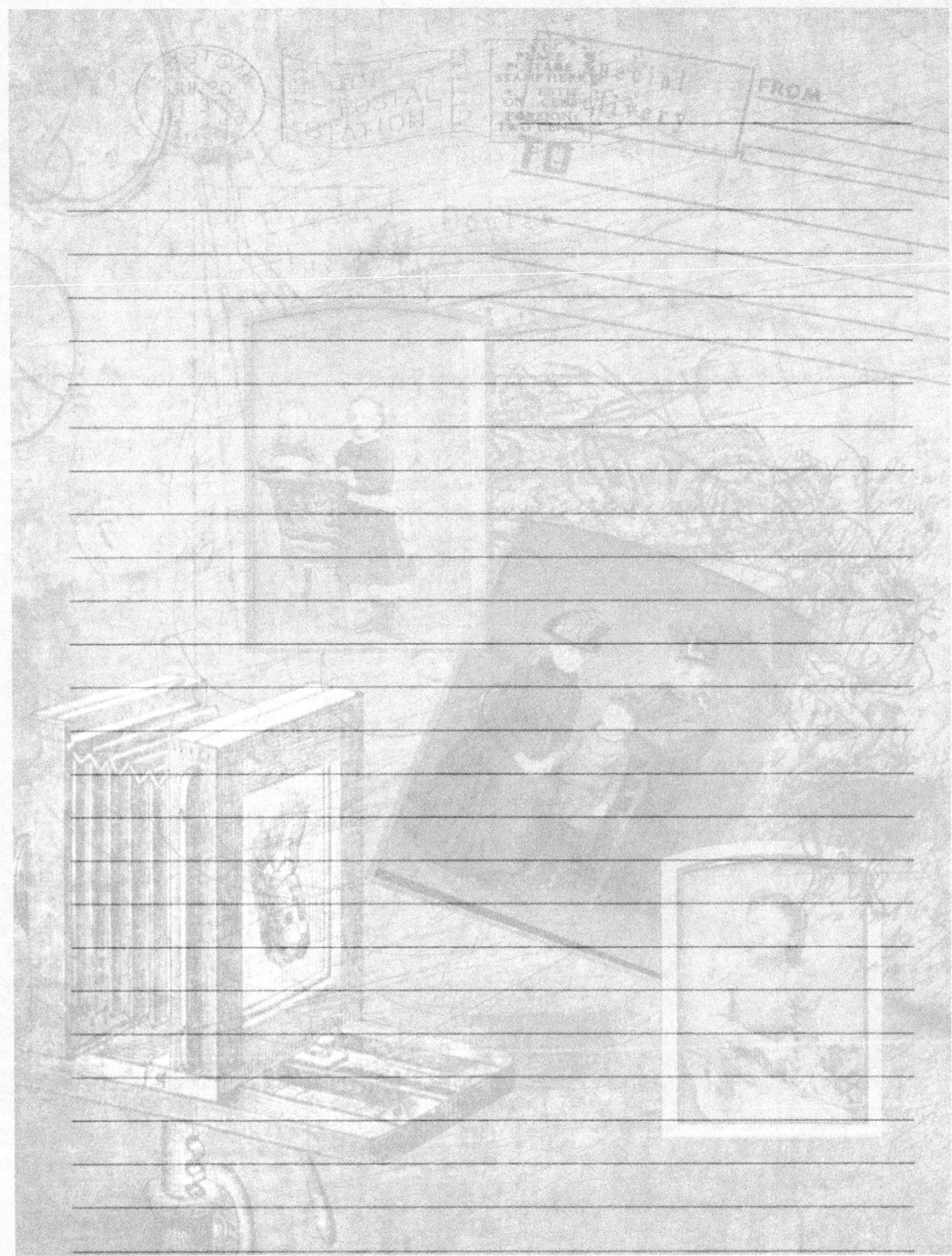

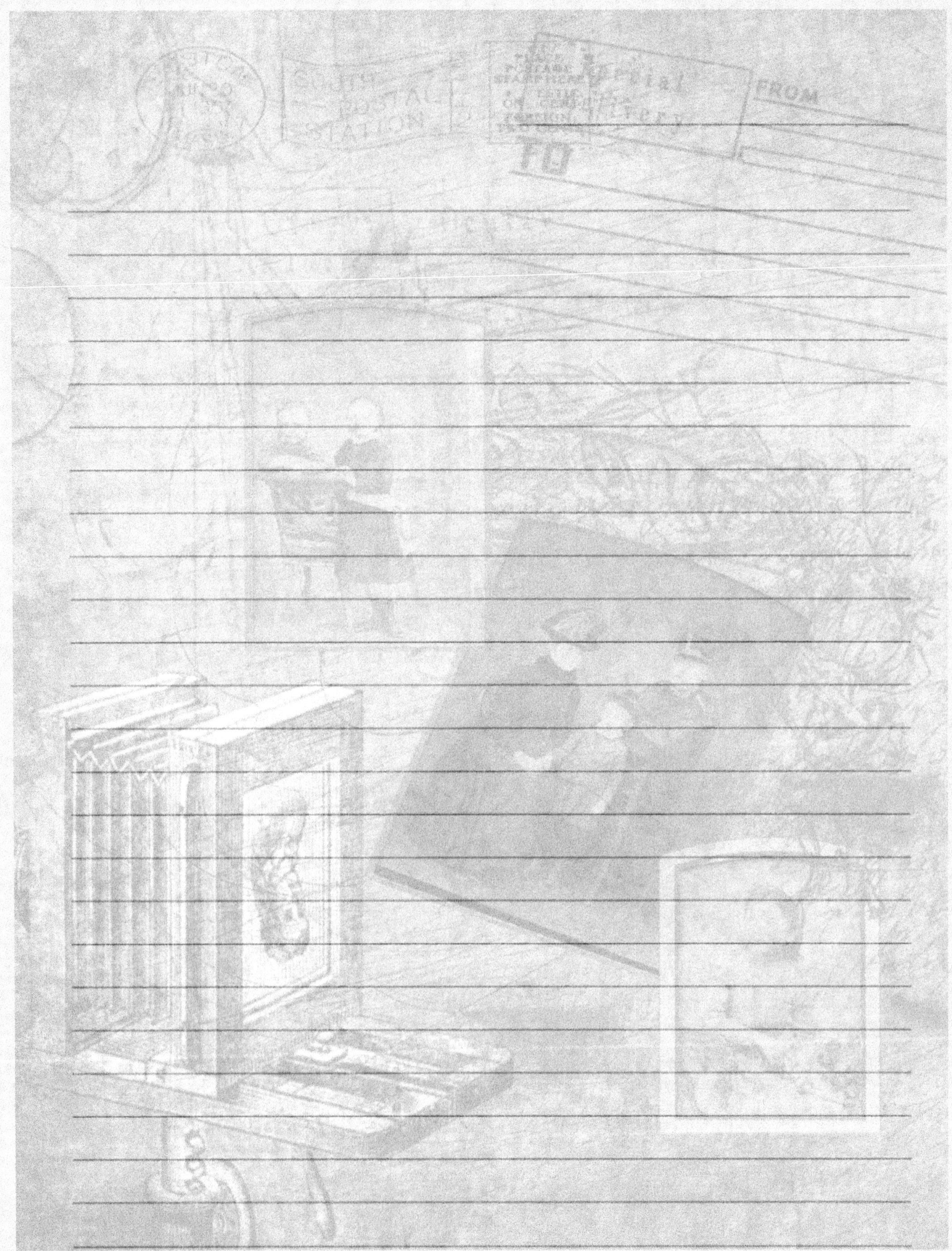

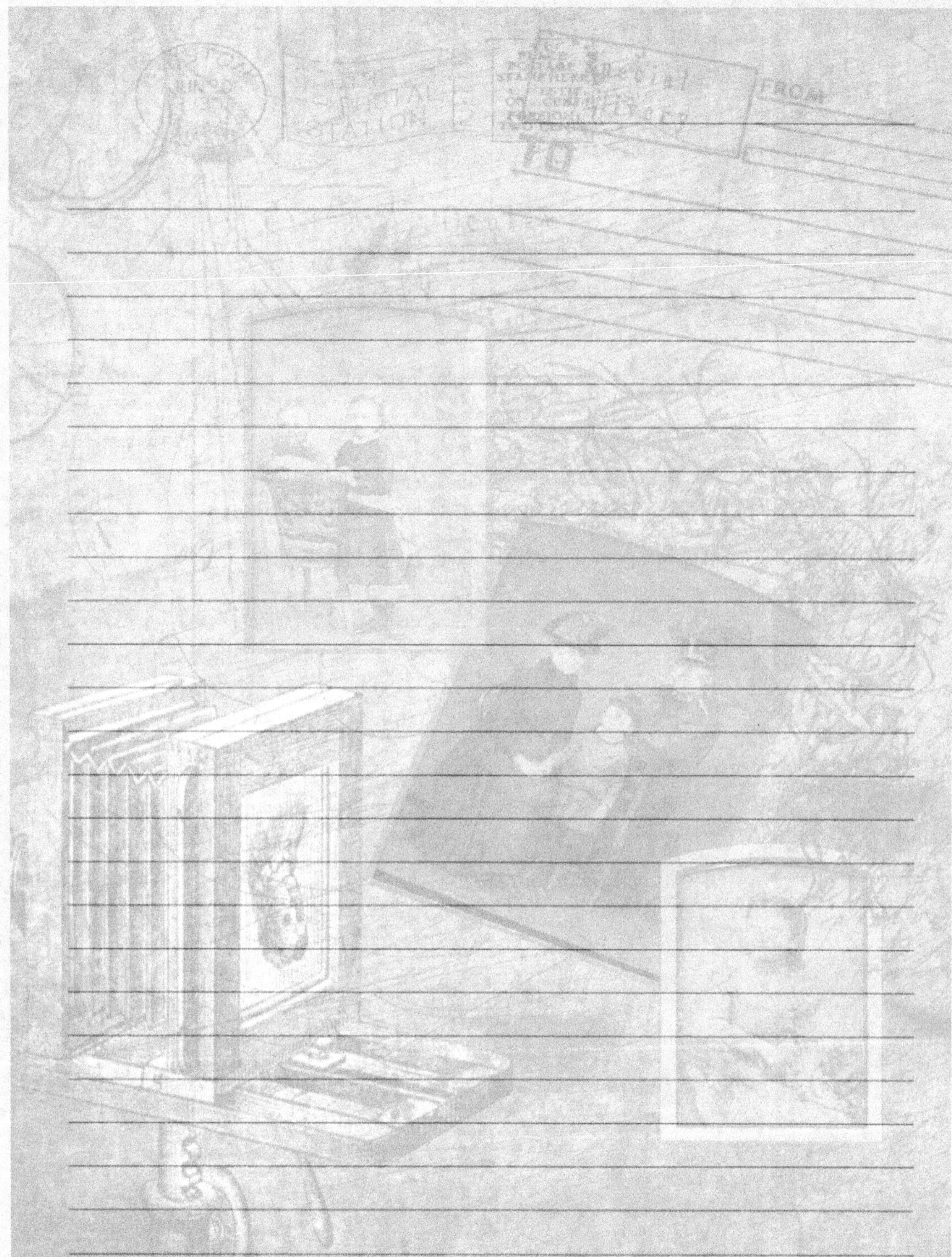

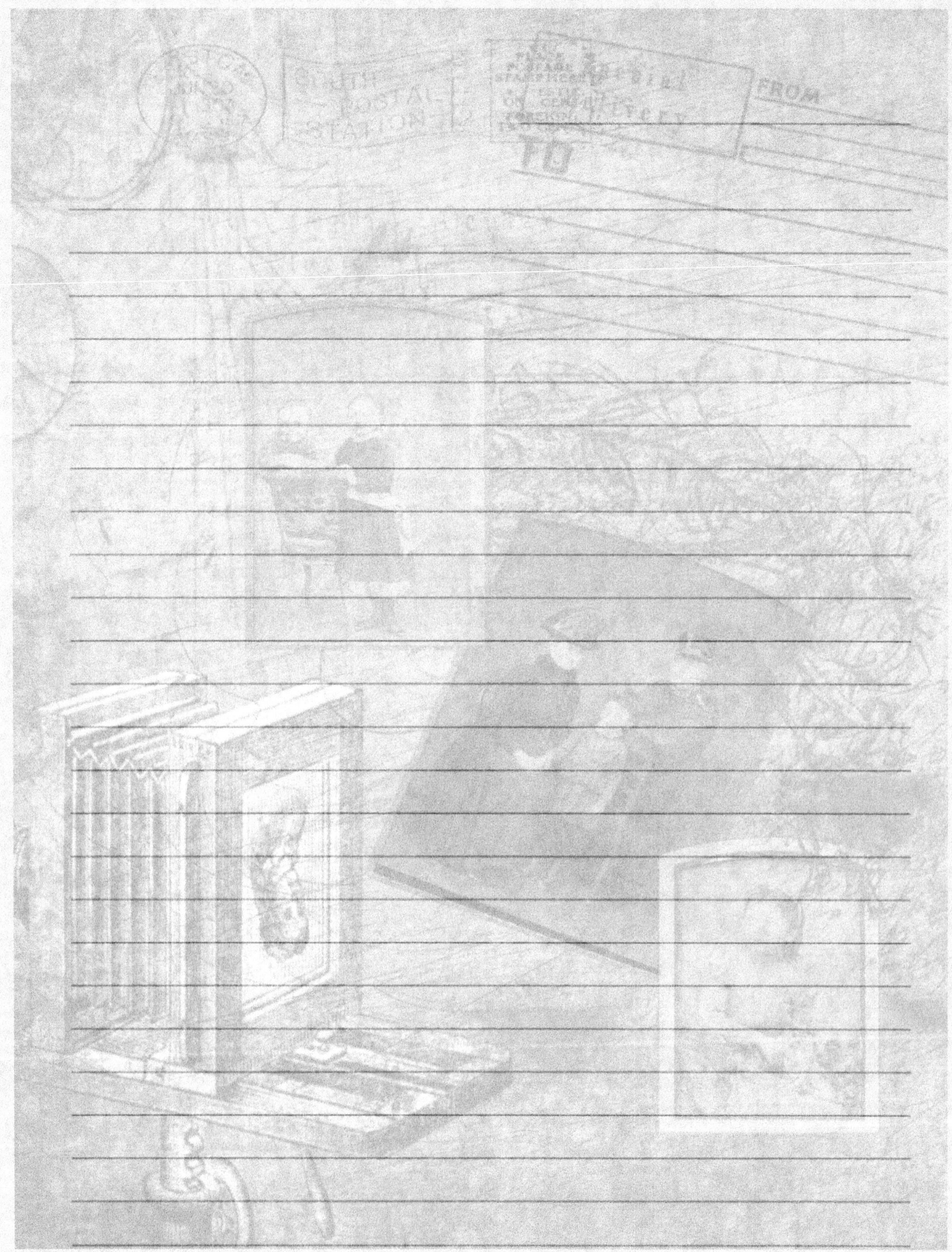

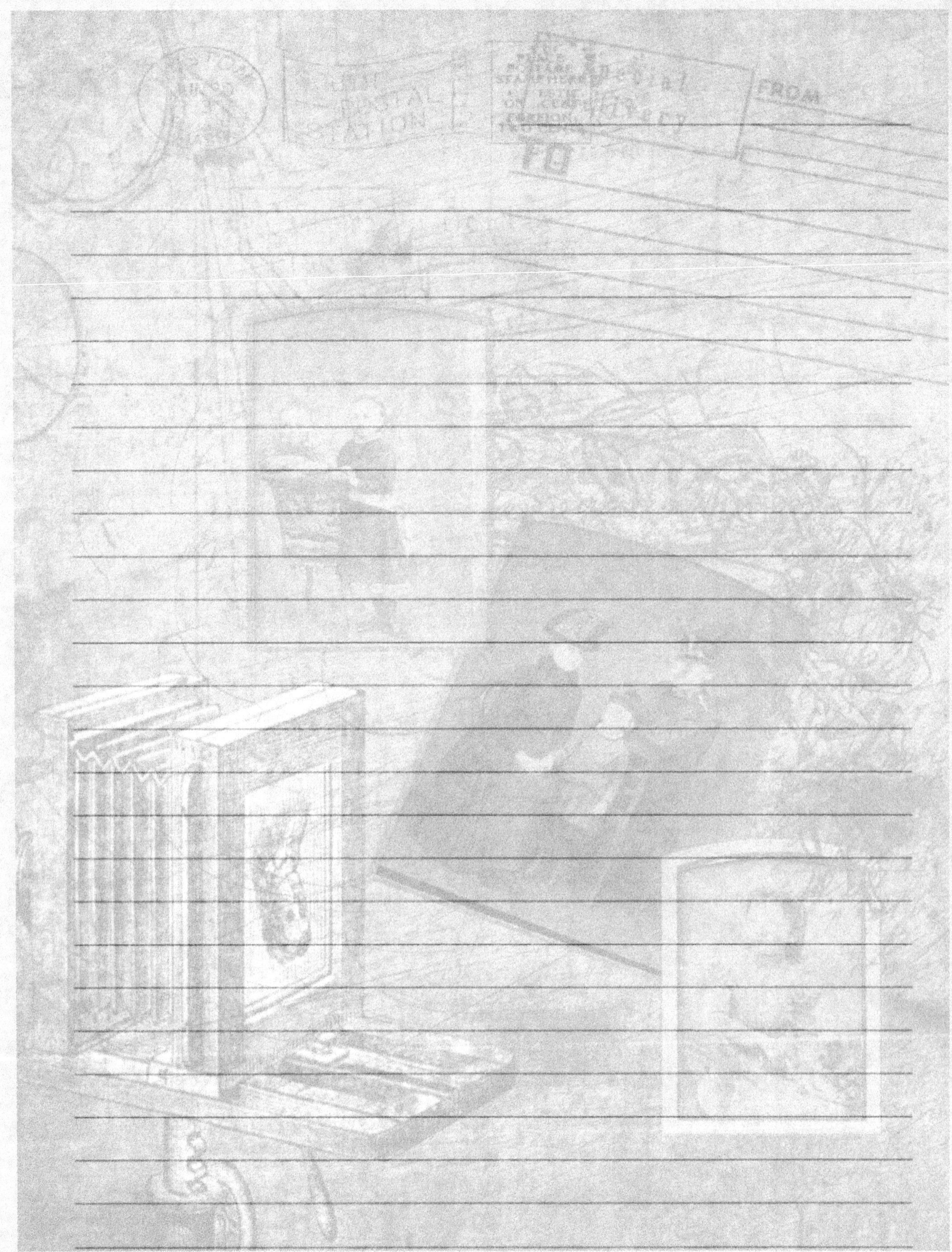

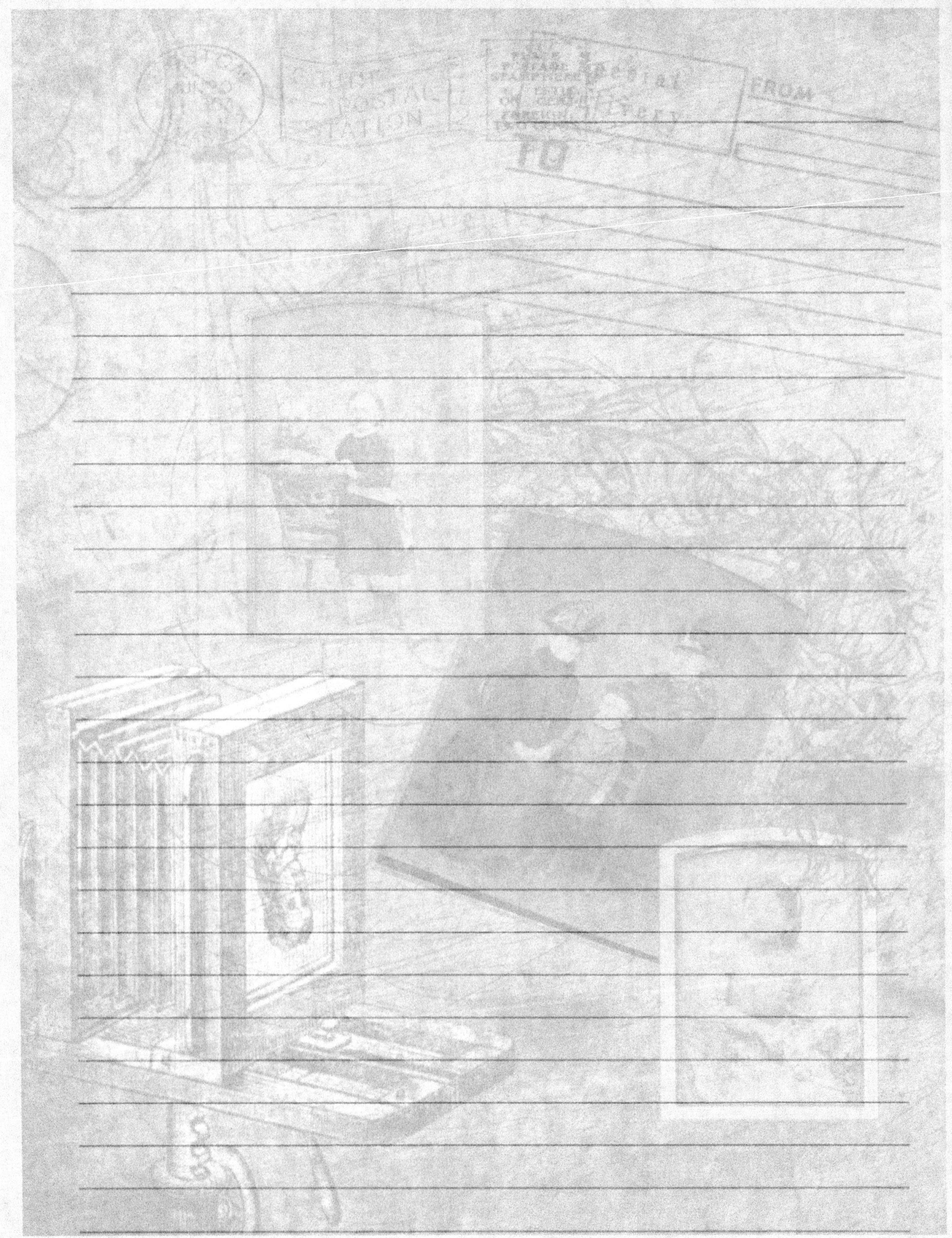

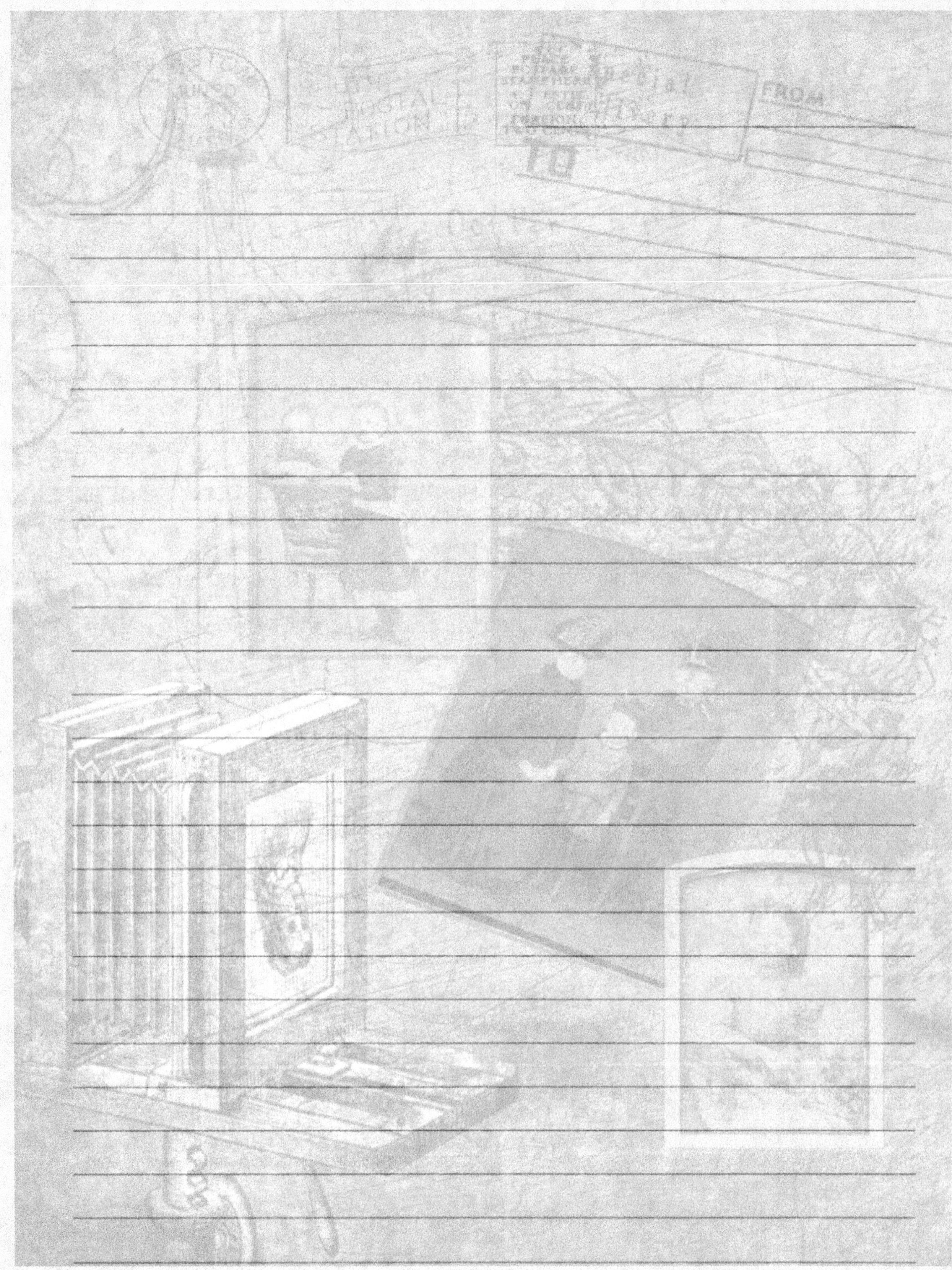

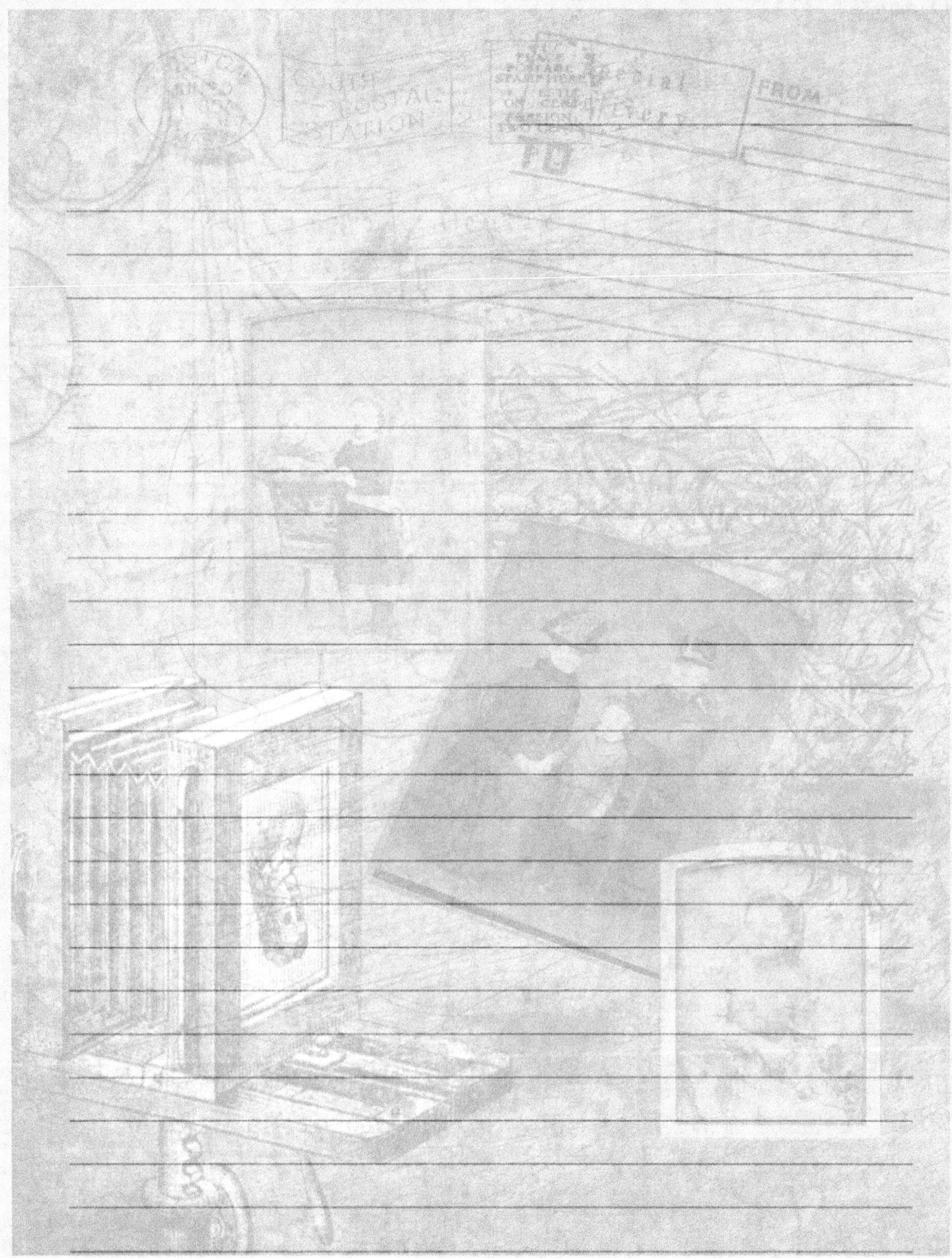

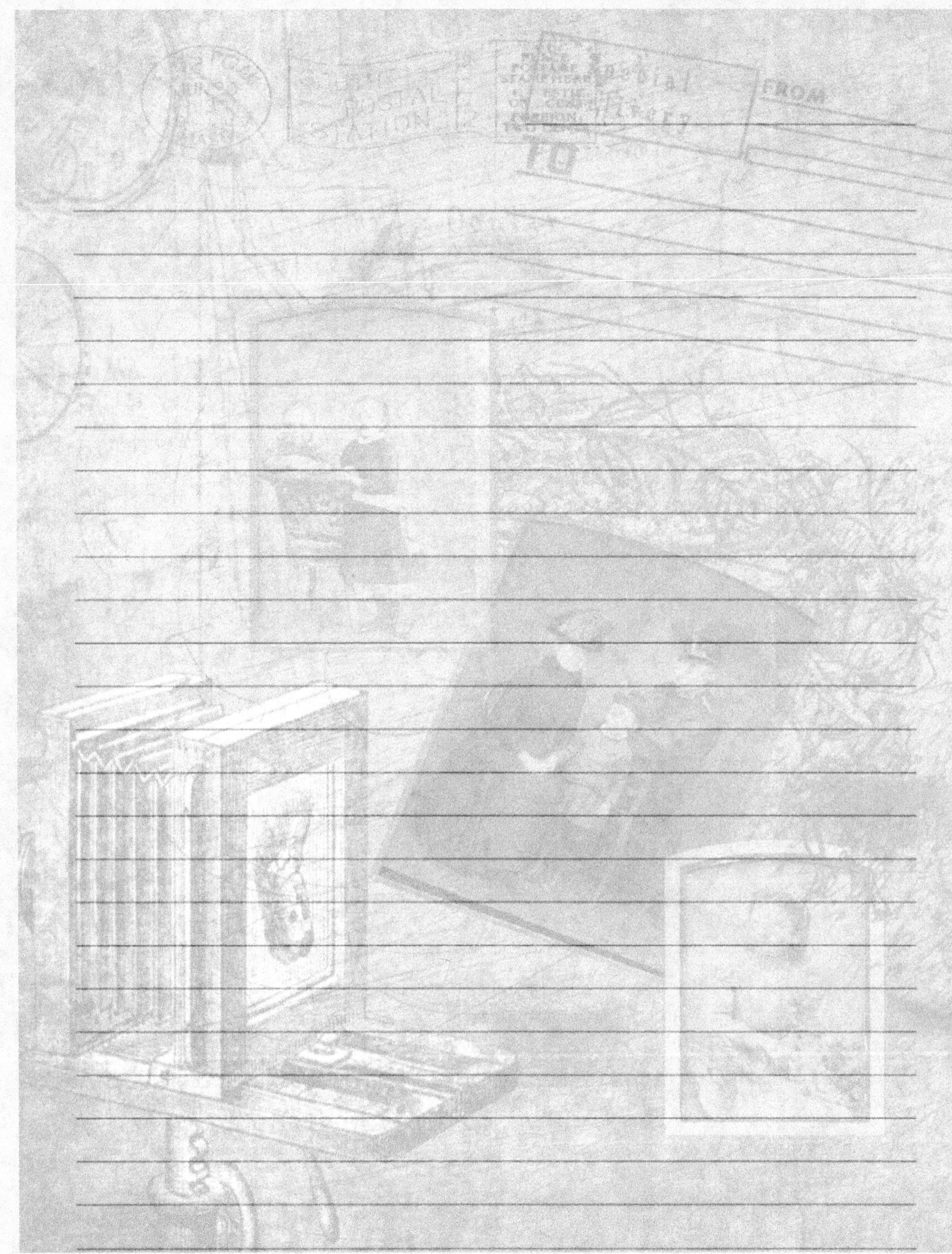

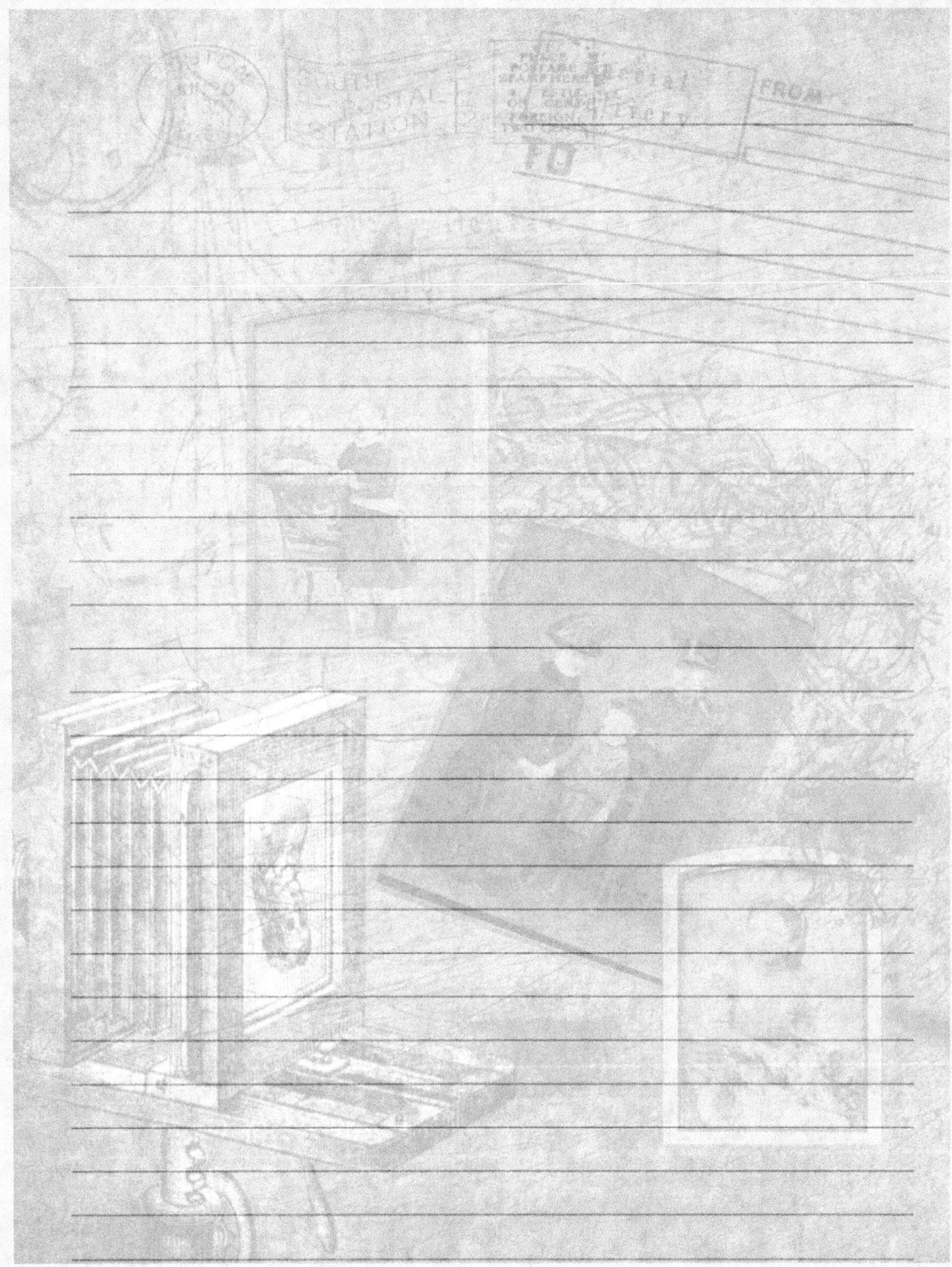

POSTAL
STATION
FROM
TO

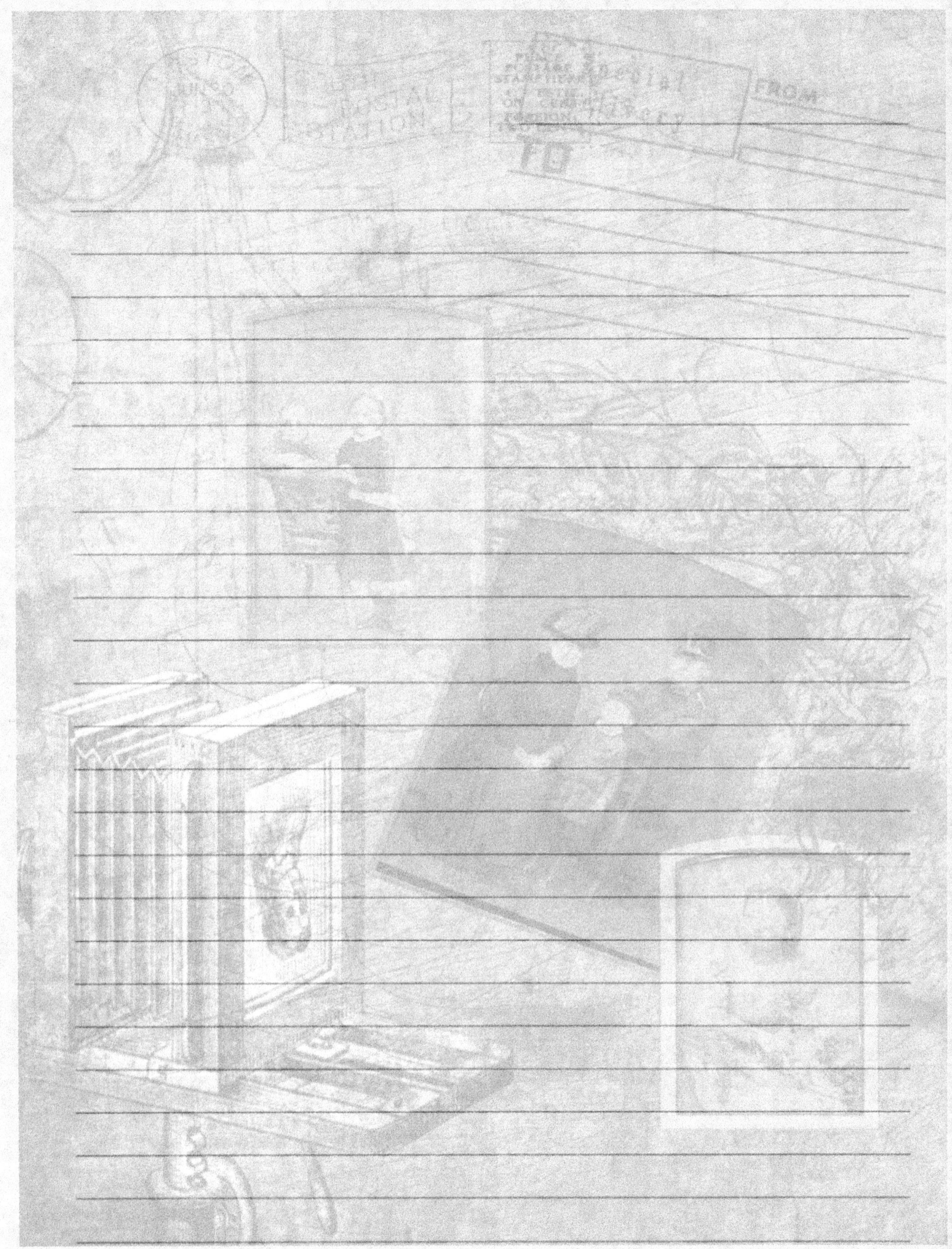

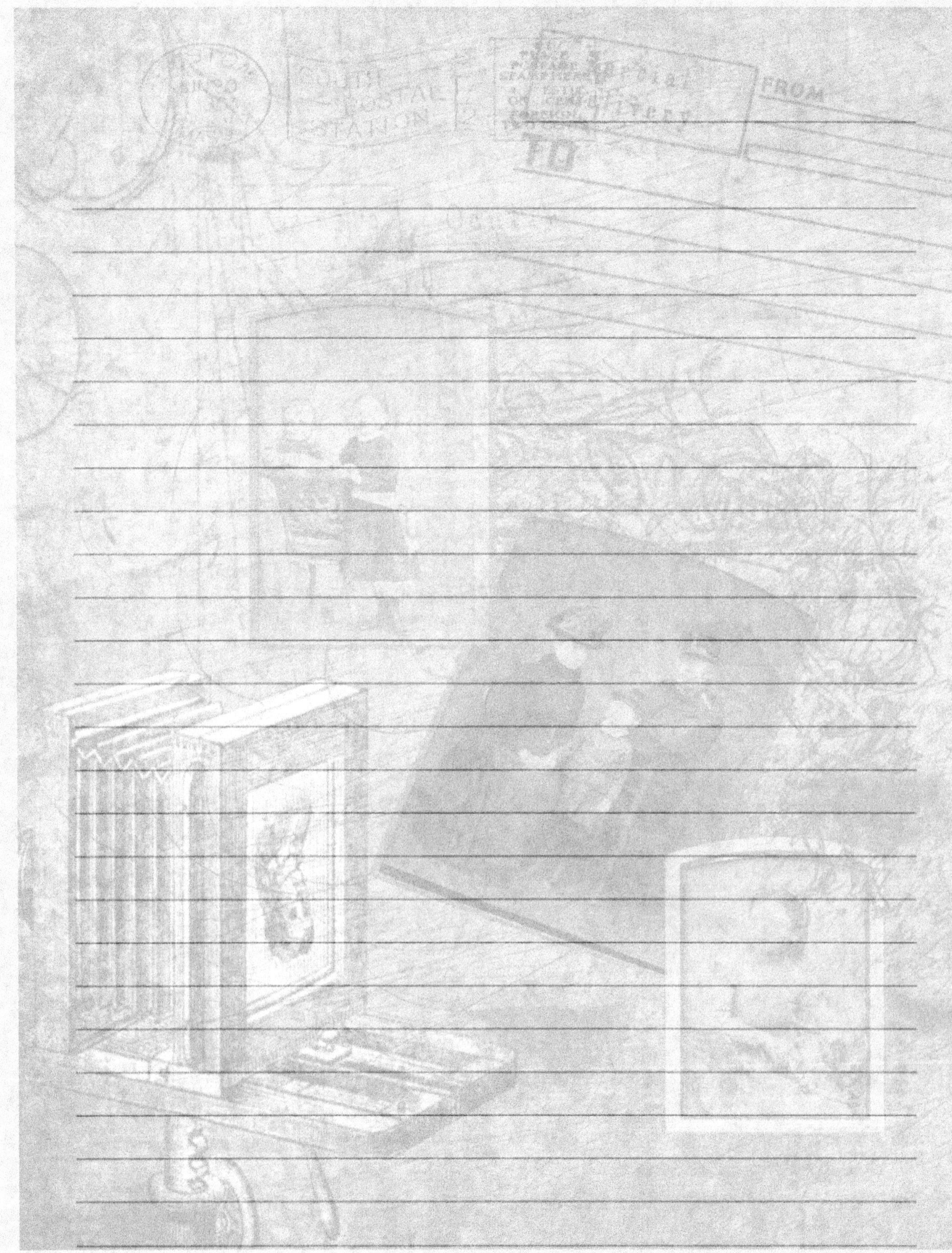

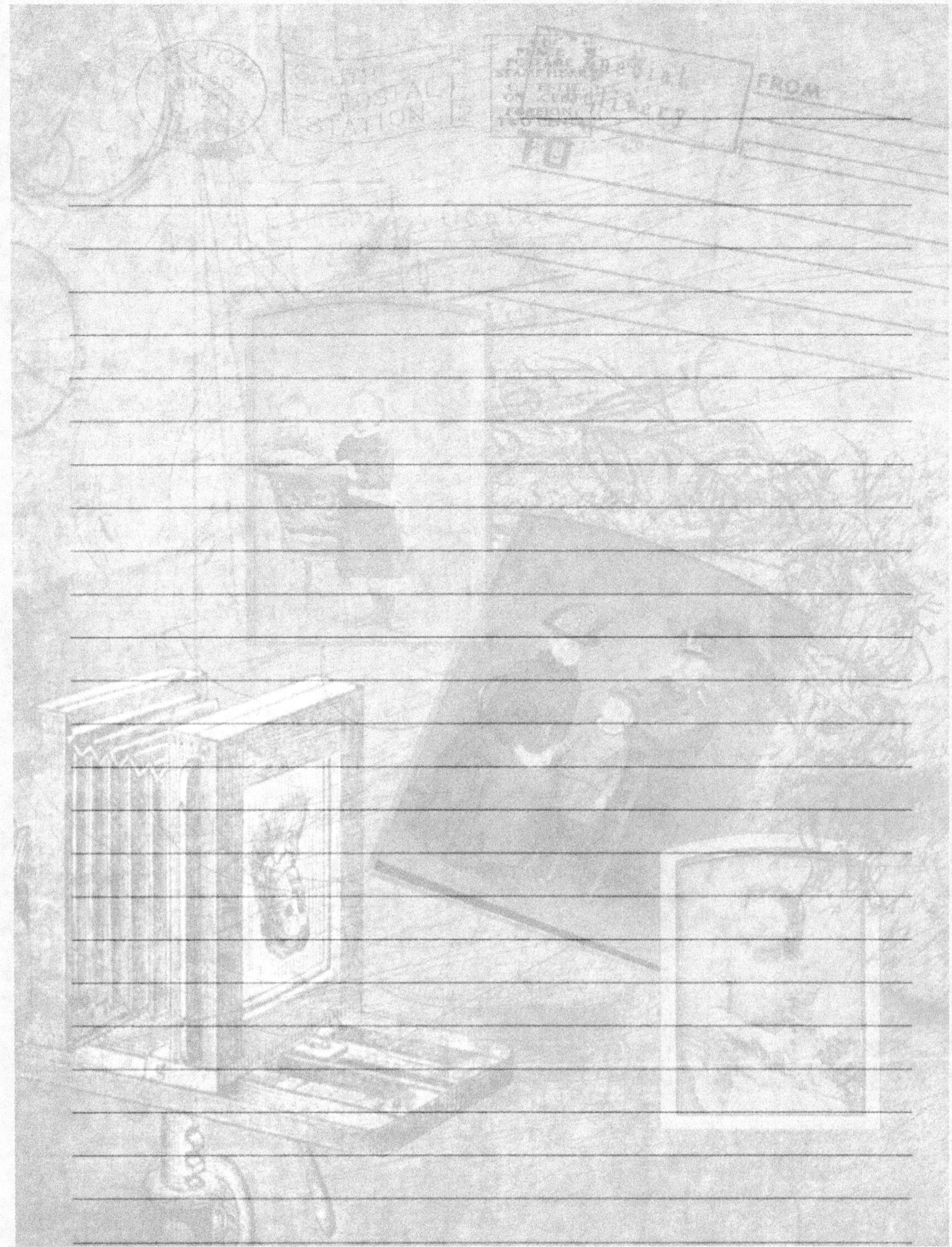

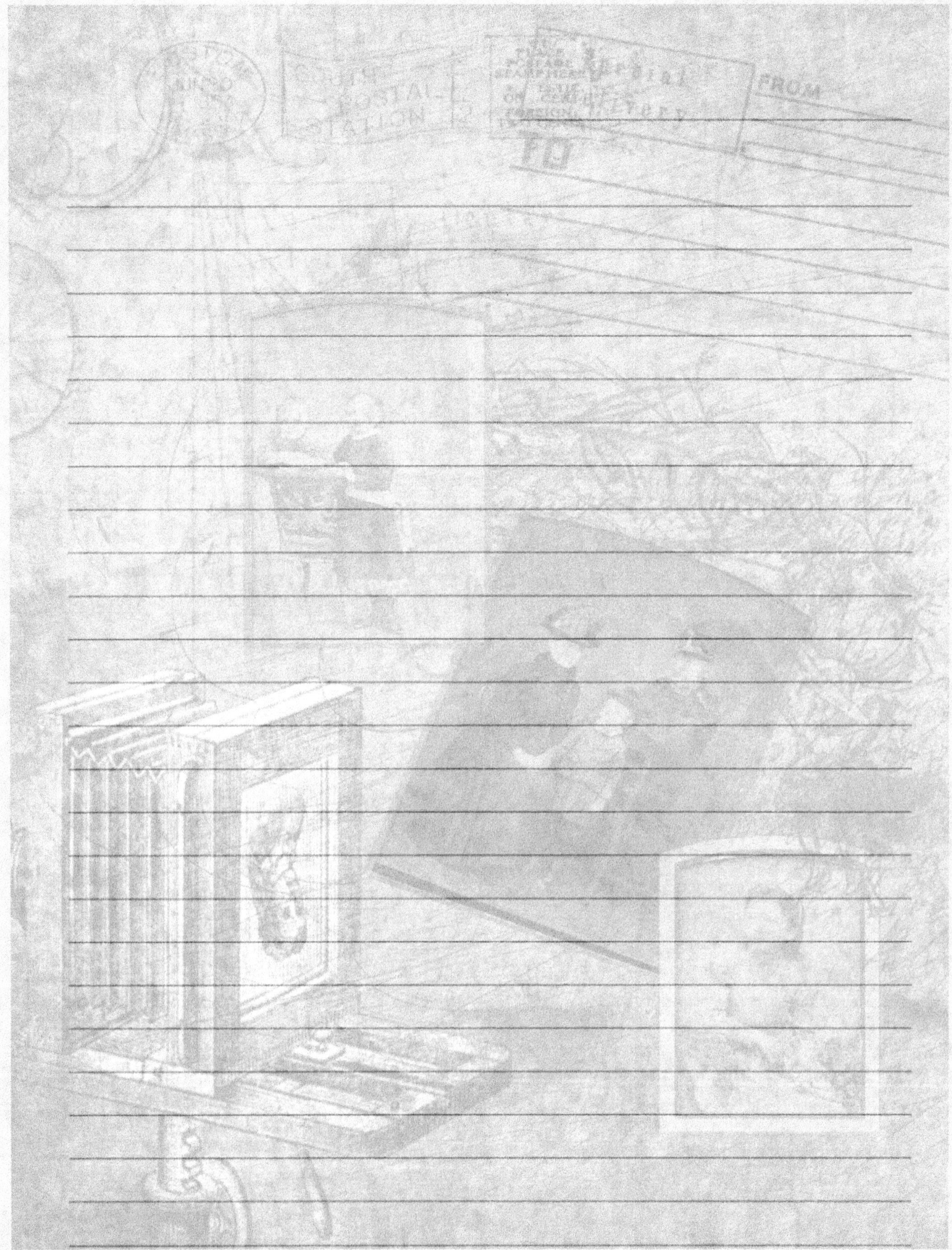

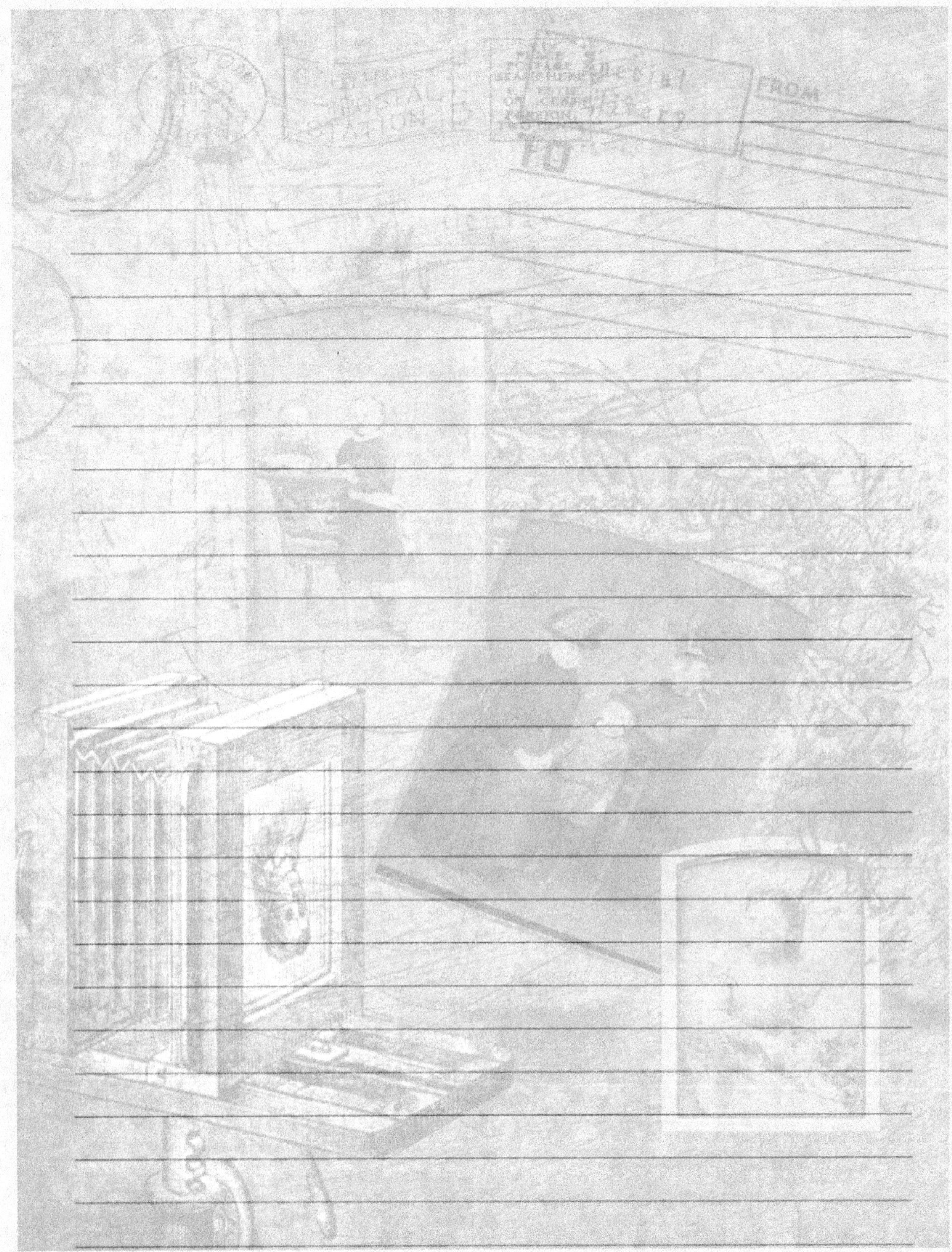

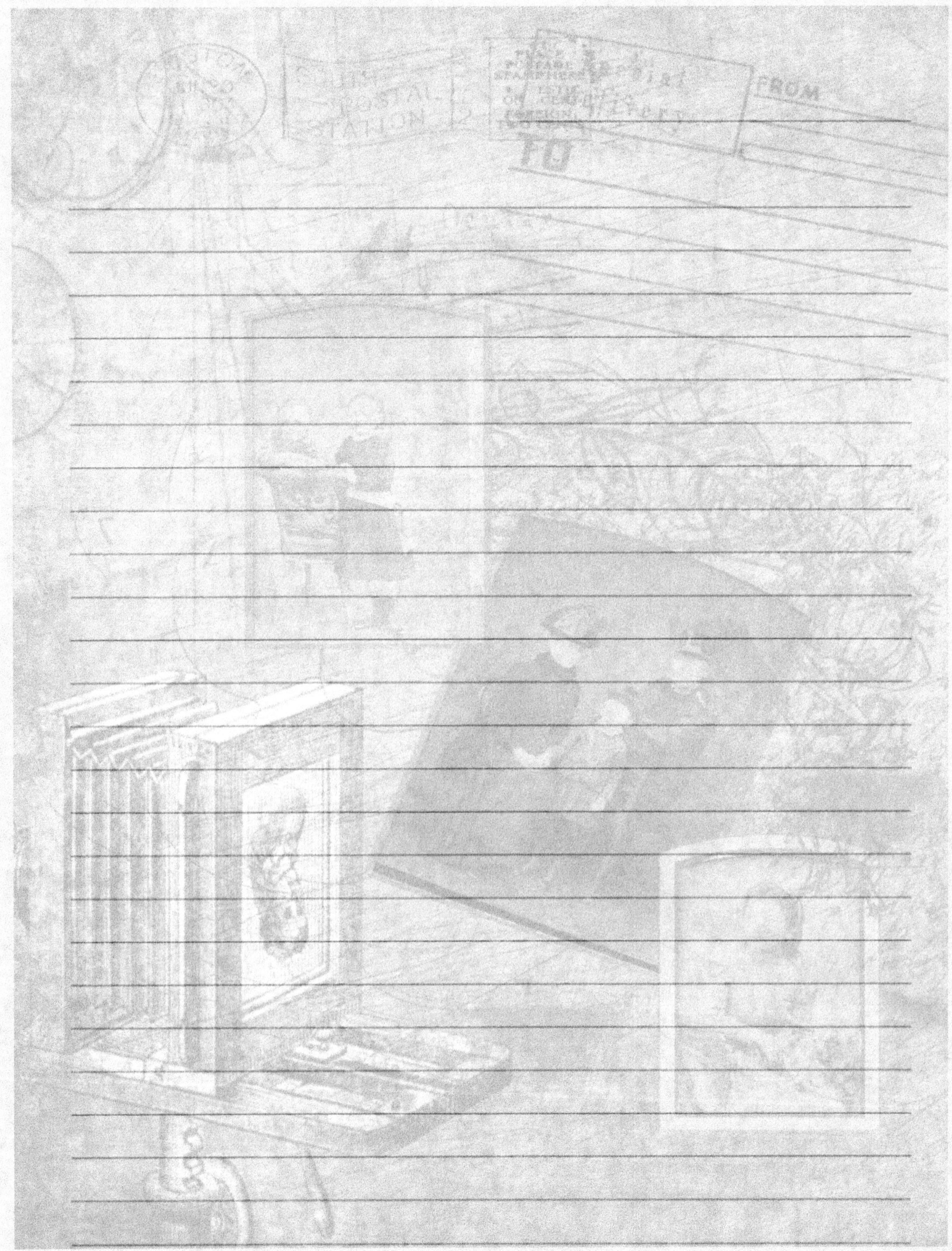
POSTAL STATION 2
SPECIAL DELIVERY
FROM
TO

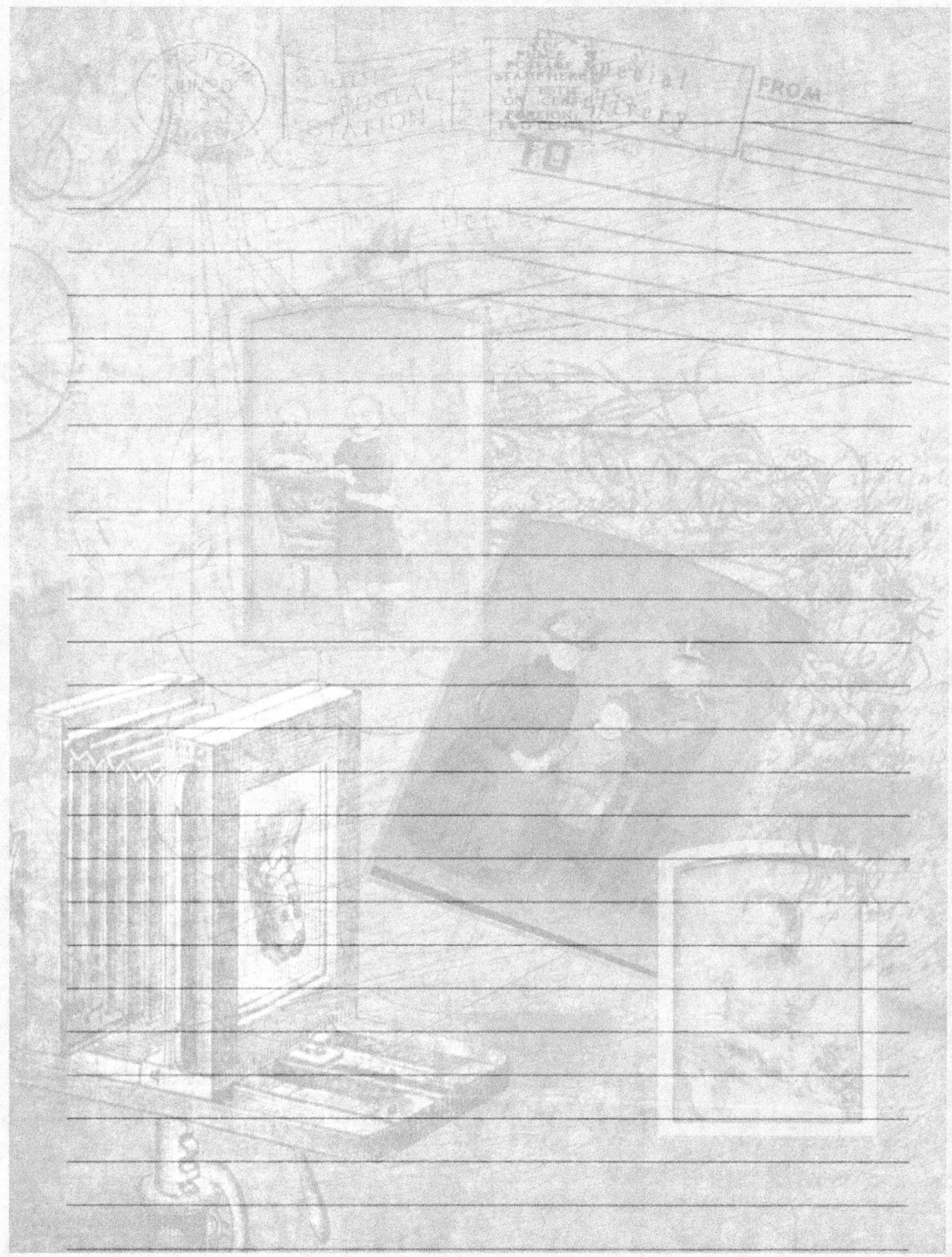

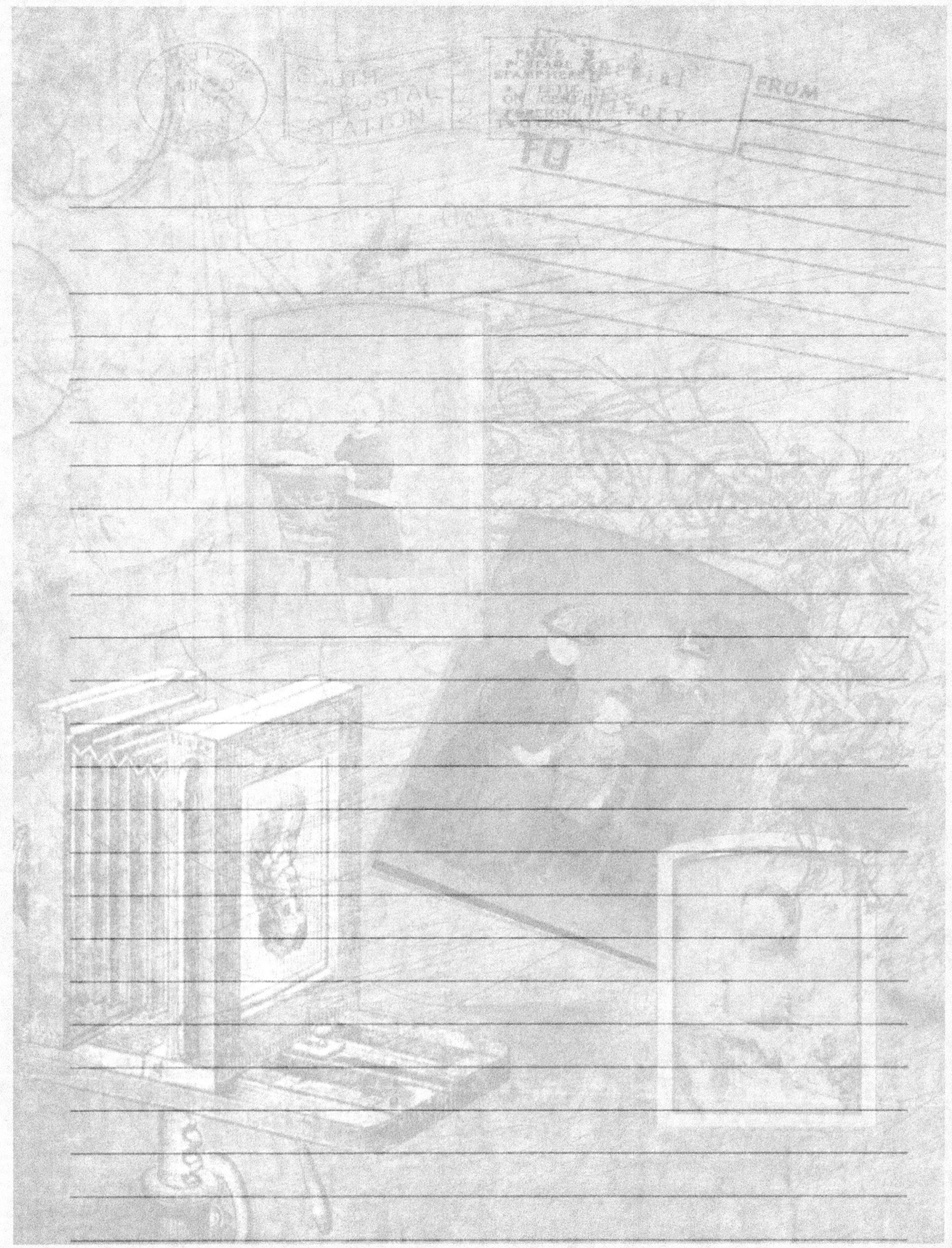
POSTAL
STATION
FROM
TO

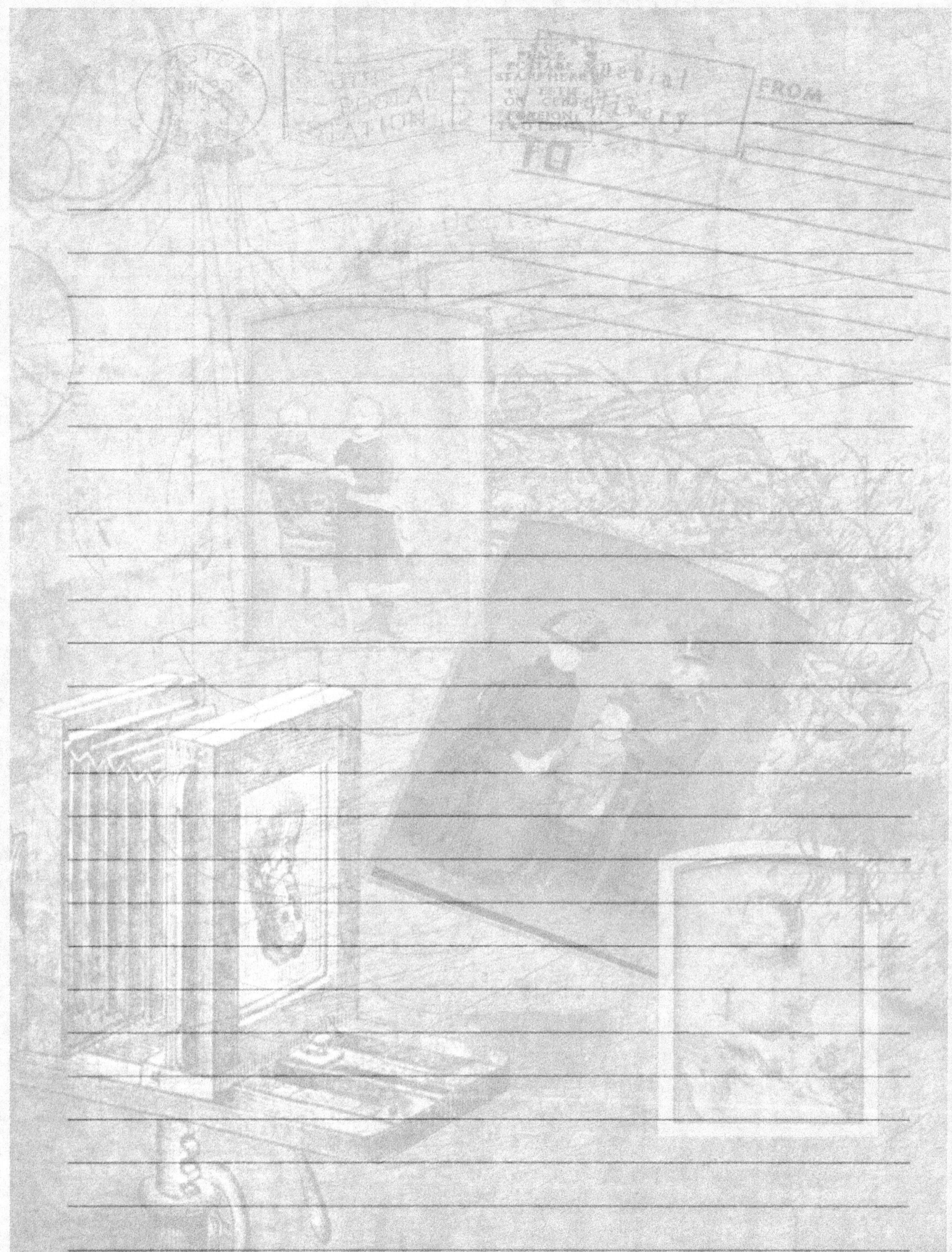

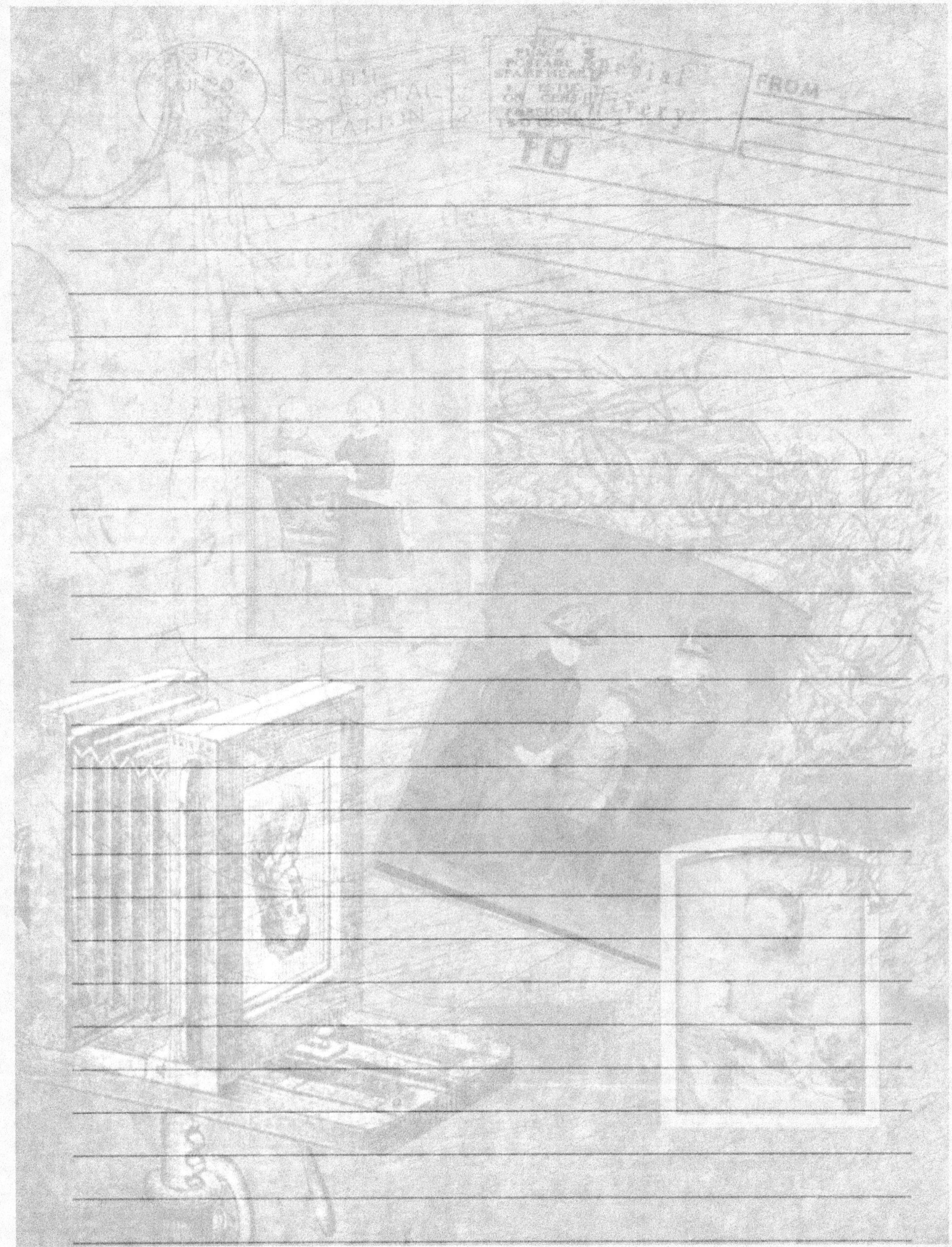

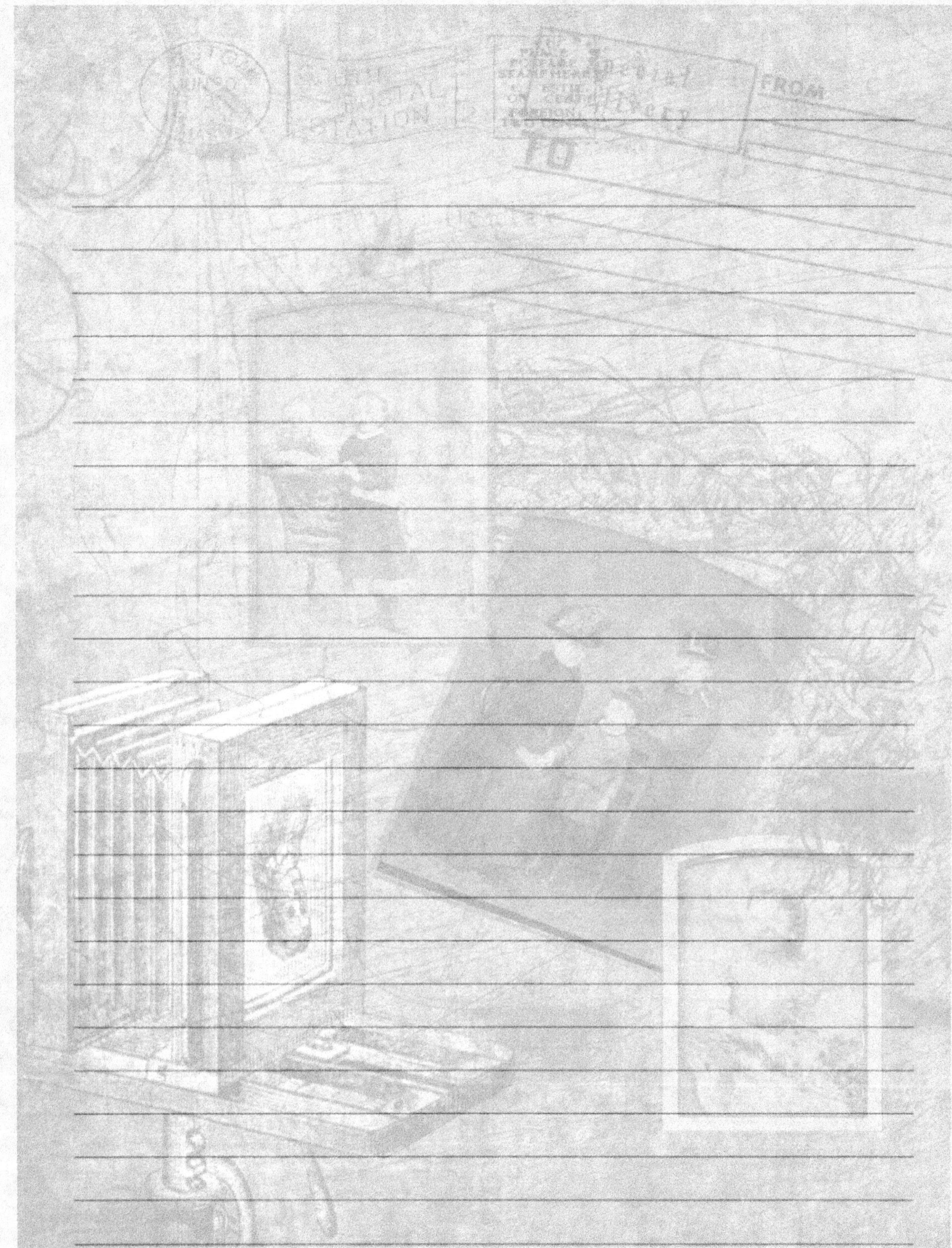

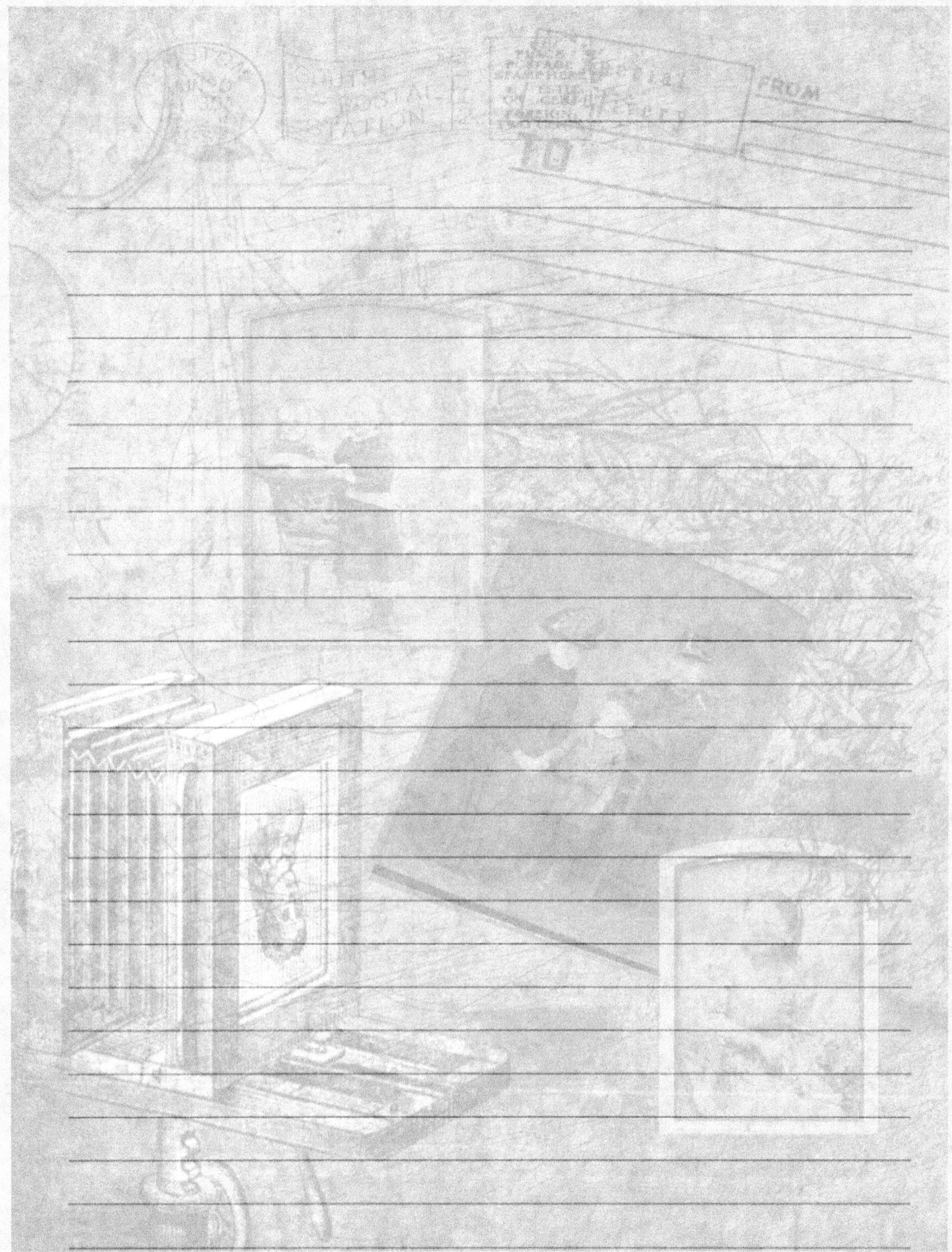
POSTAL STATION
SPECIAL DELIVERY
FROM
TO

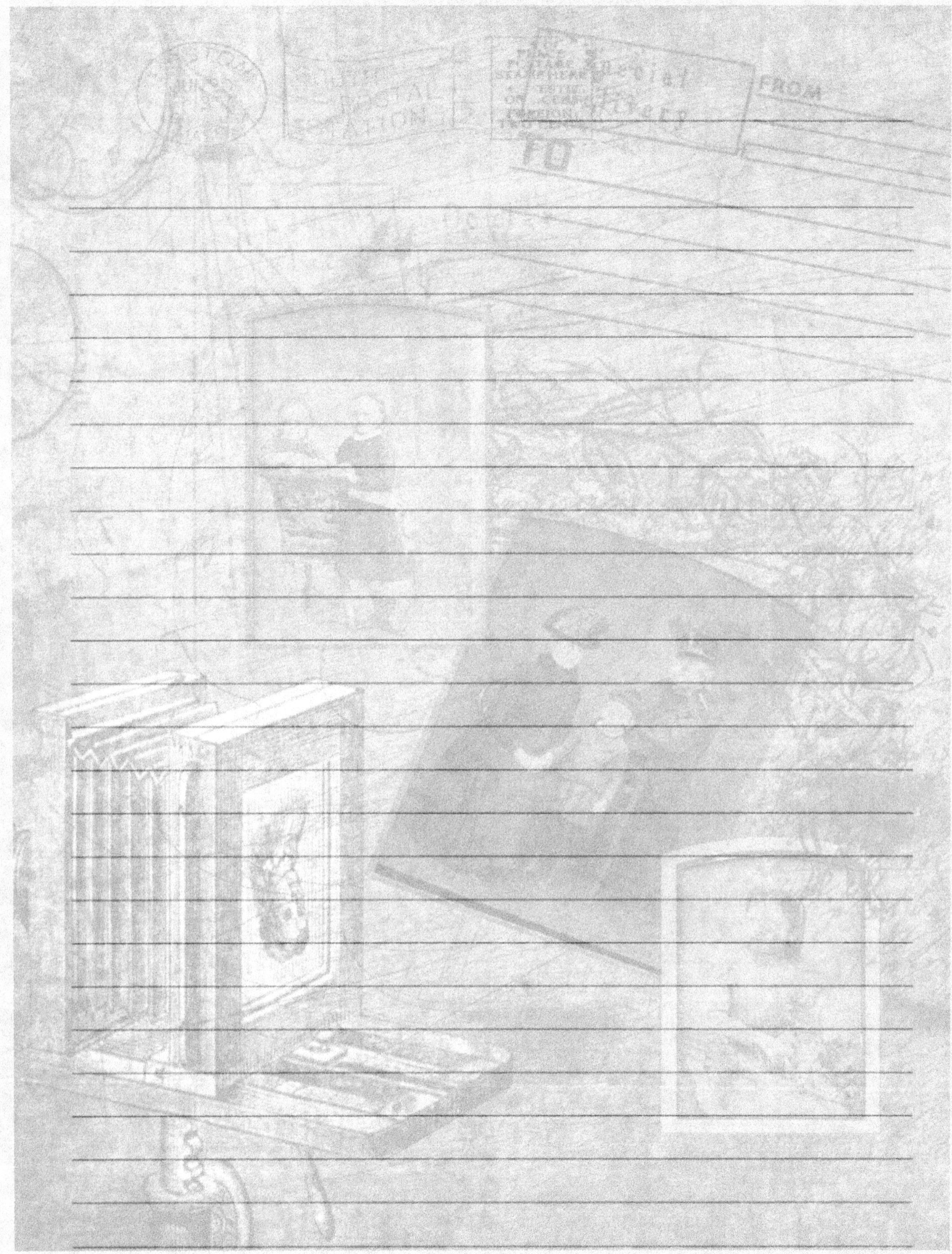

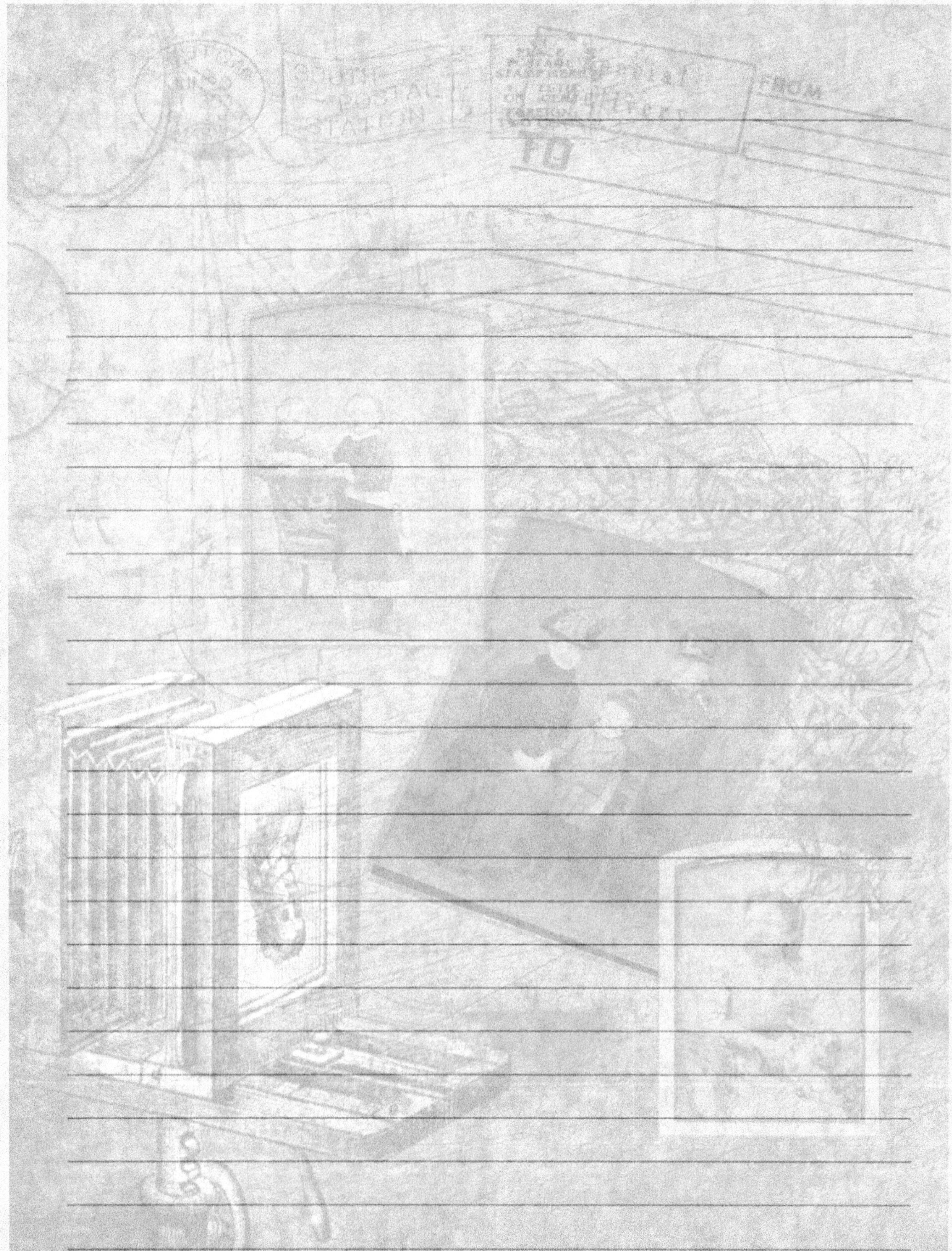

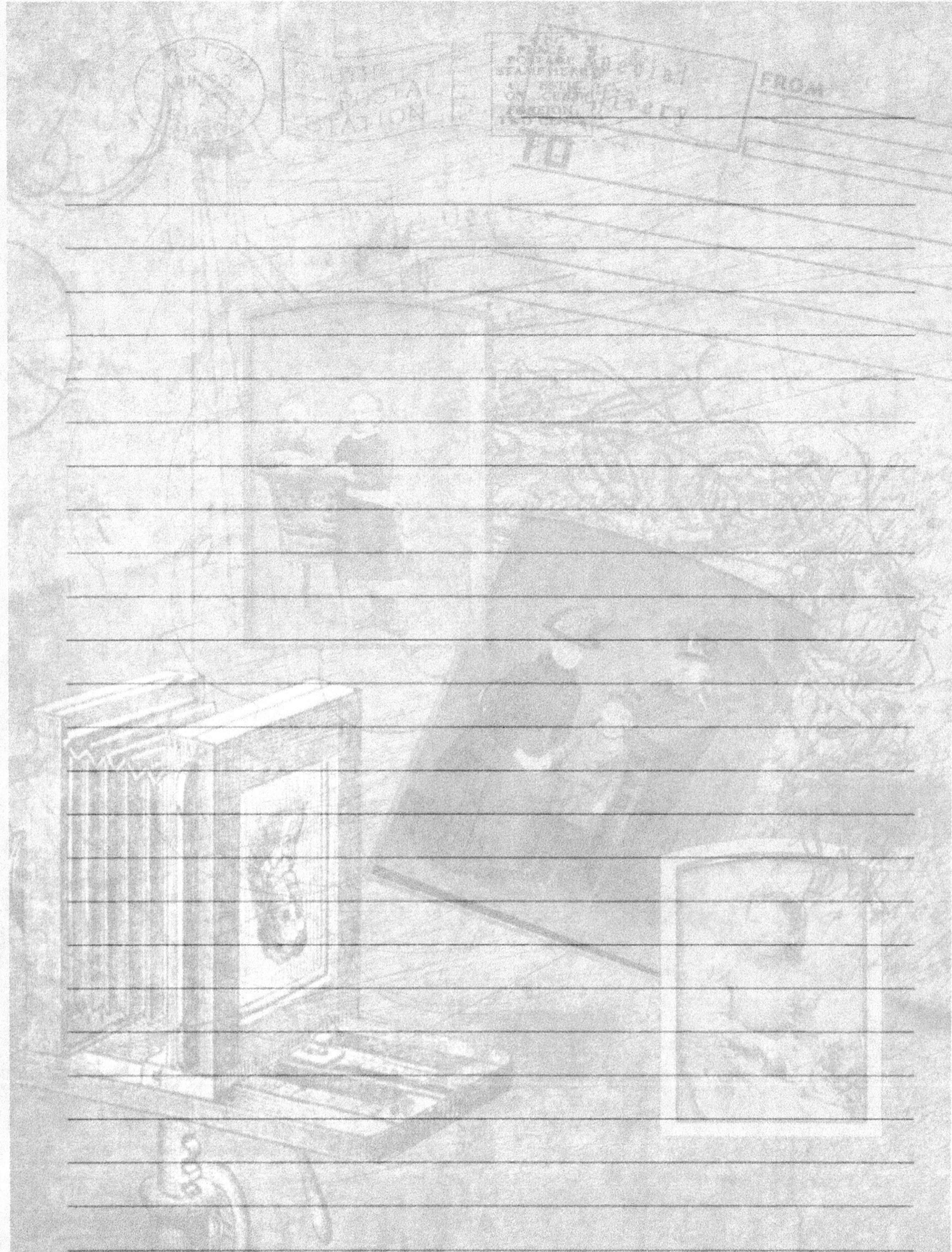

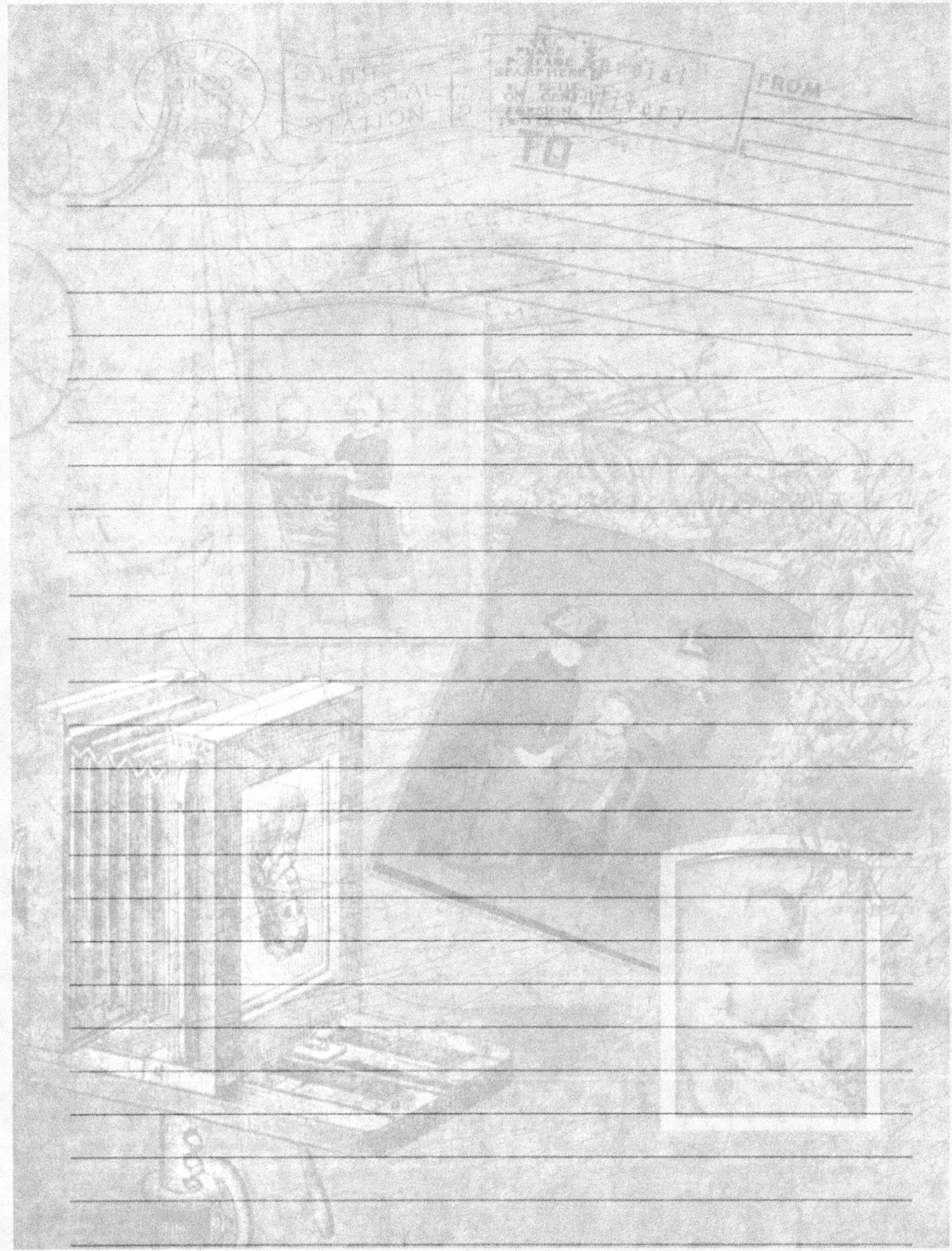

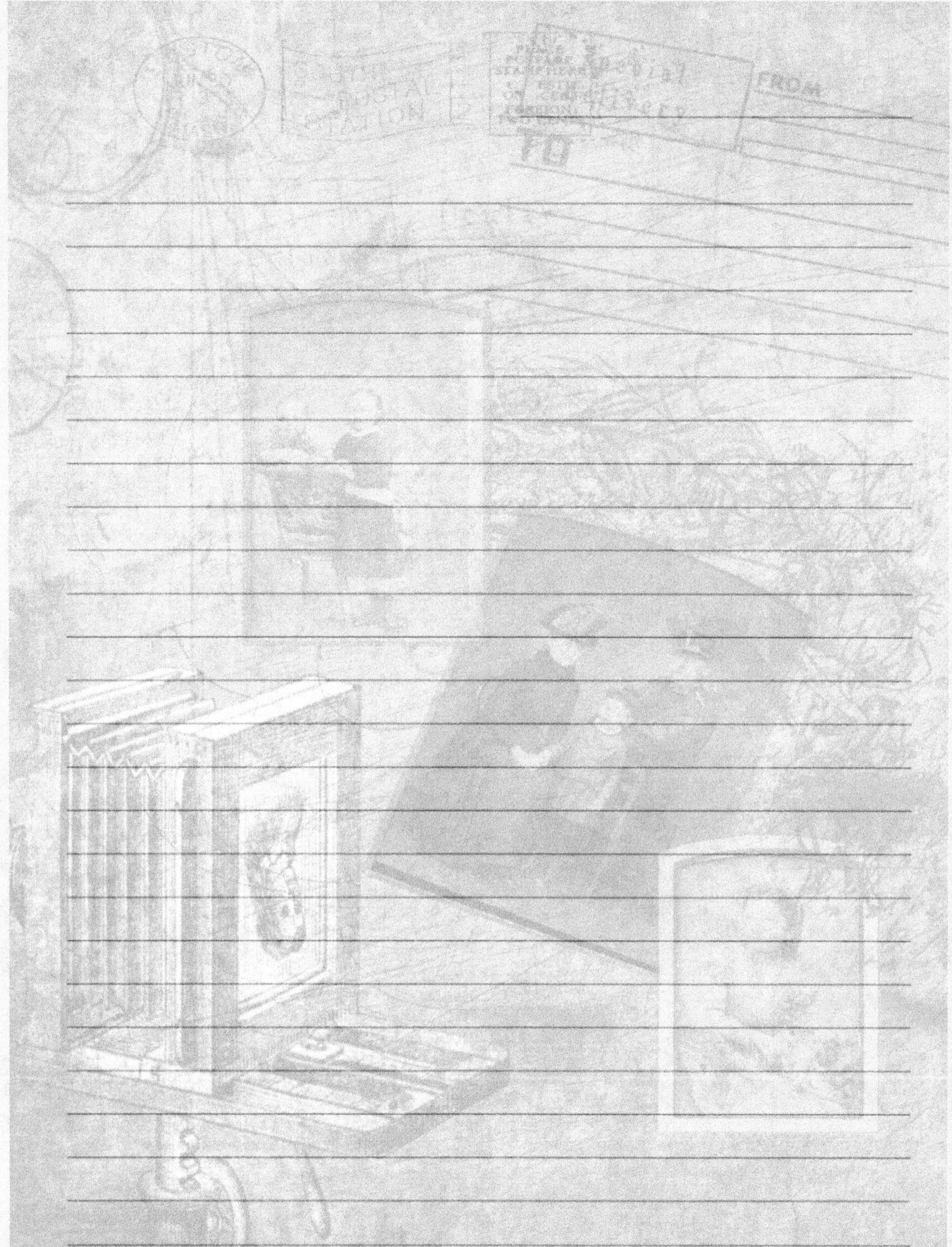

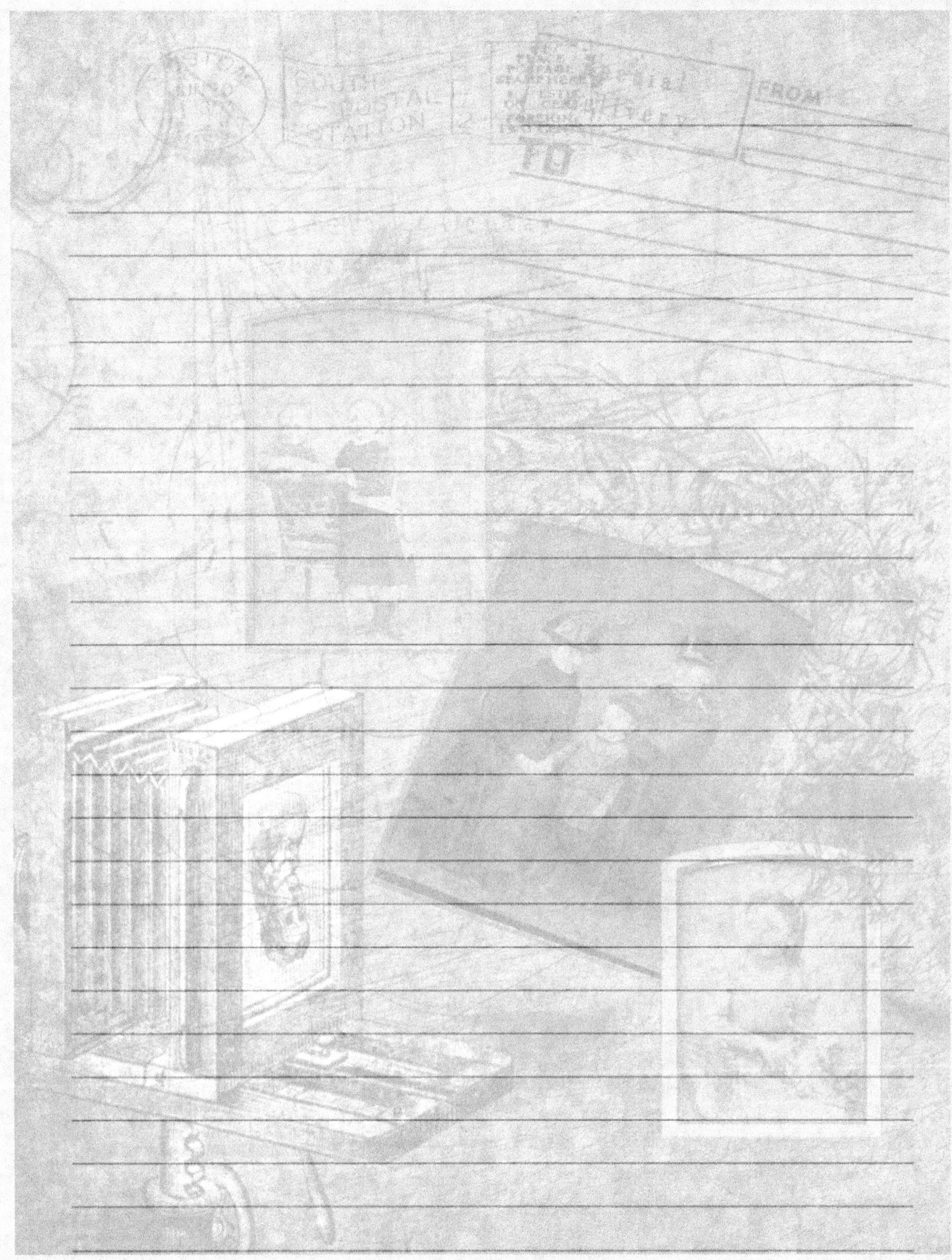

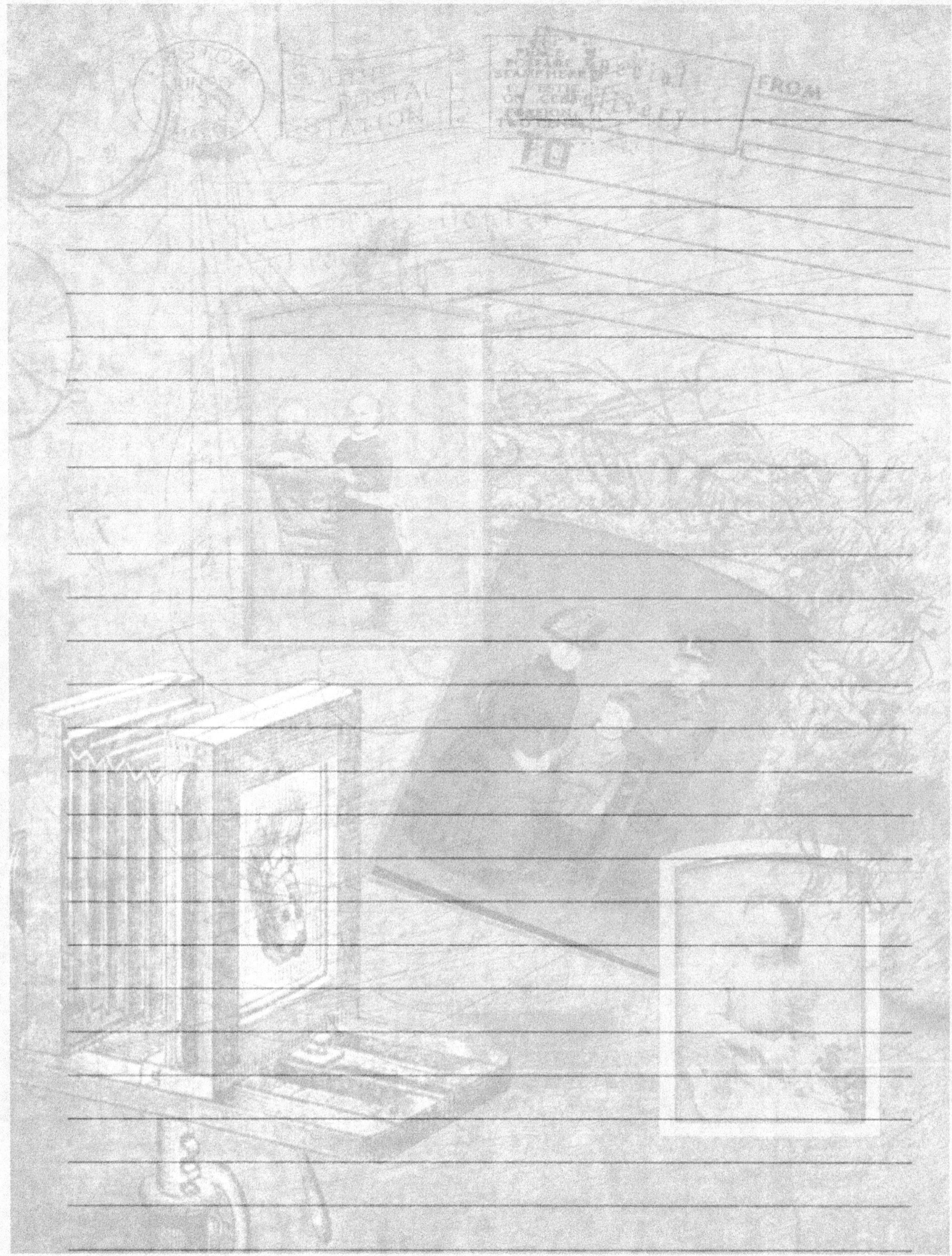

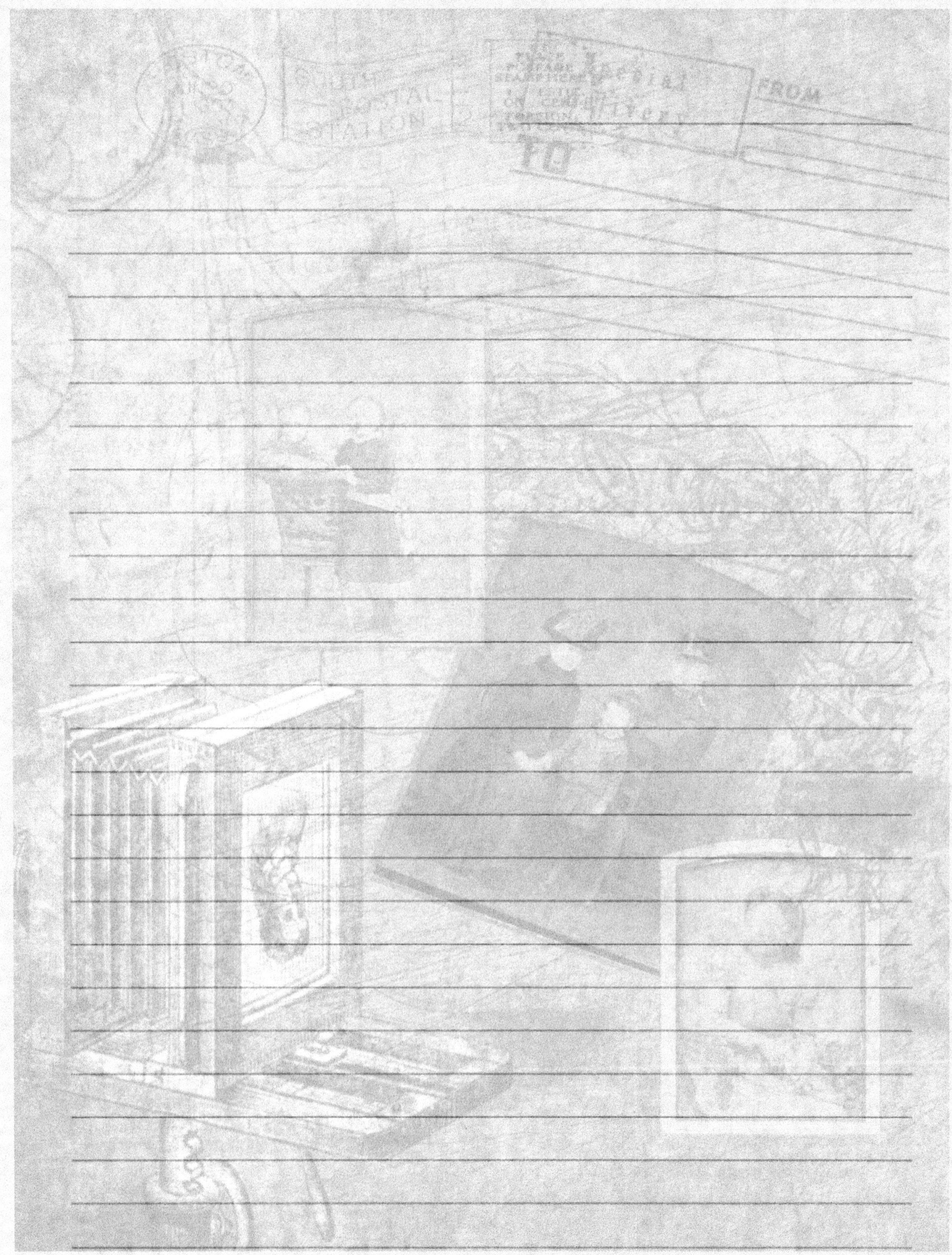

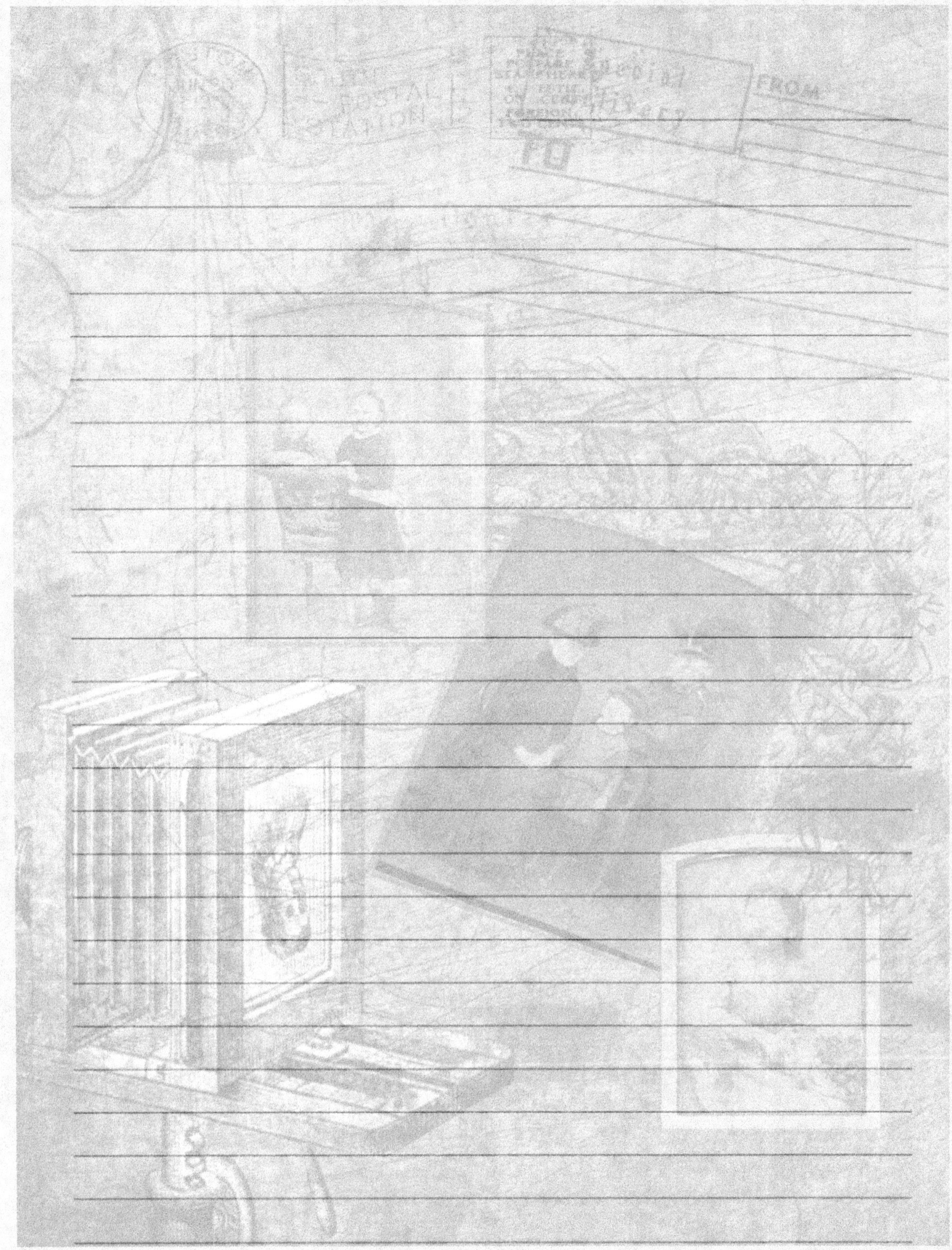

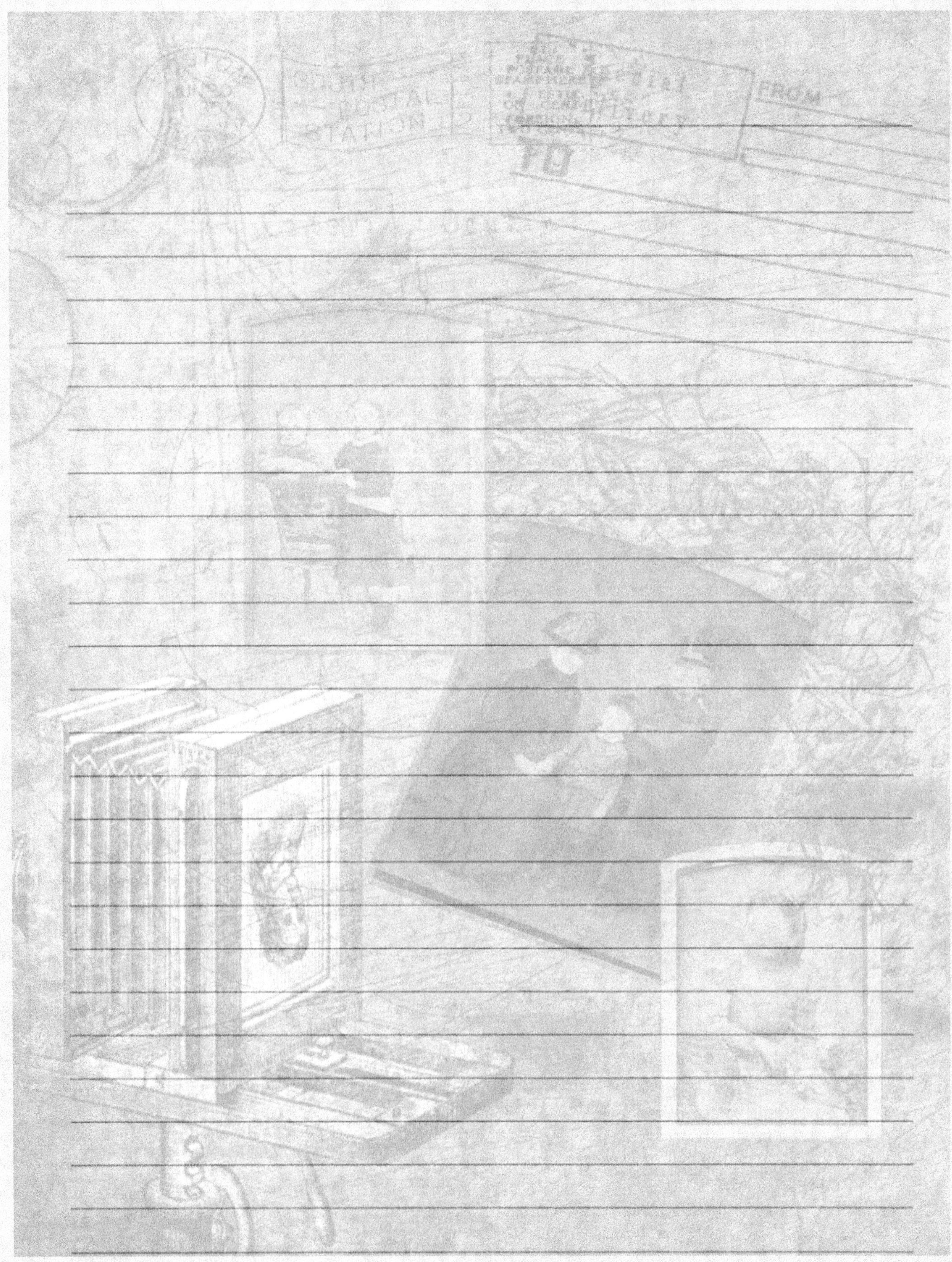

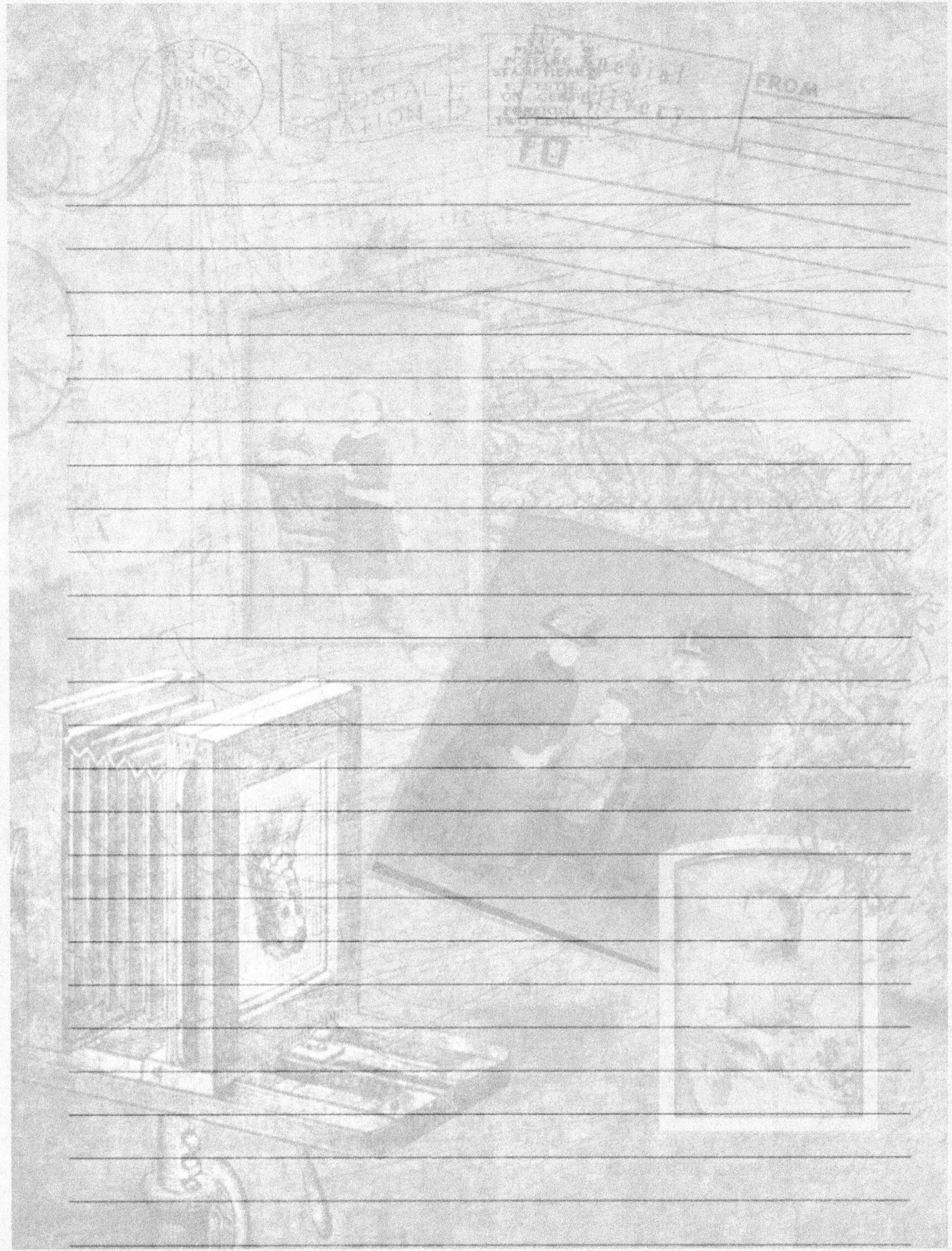

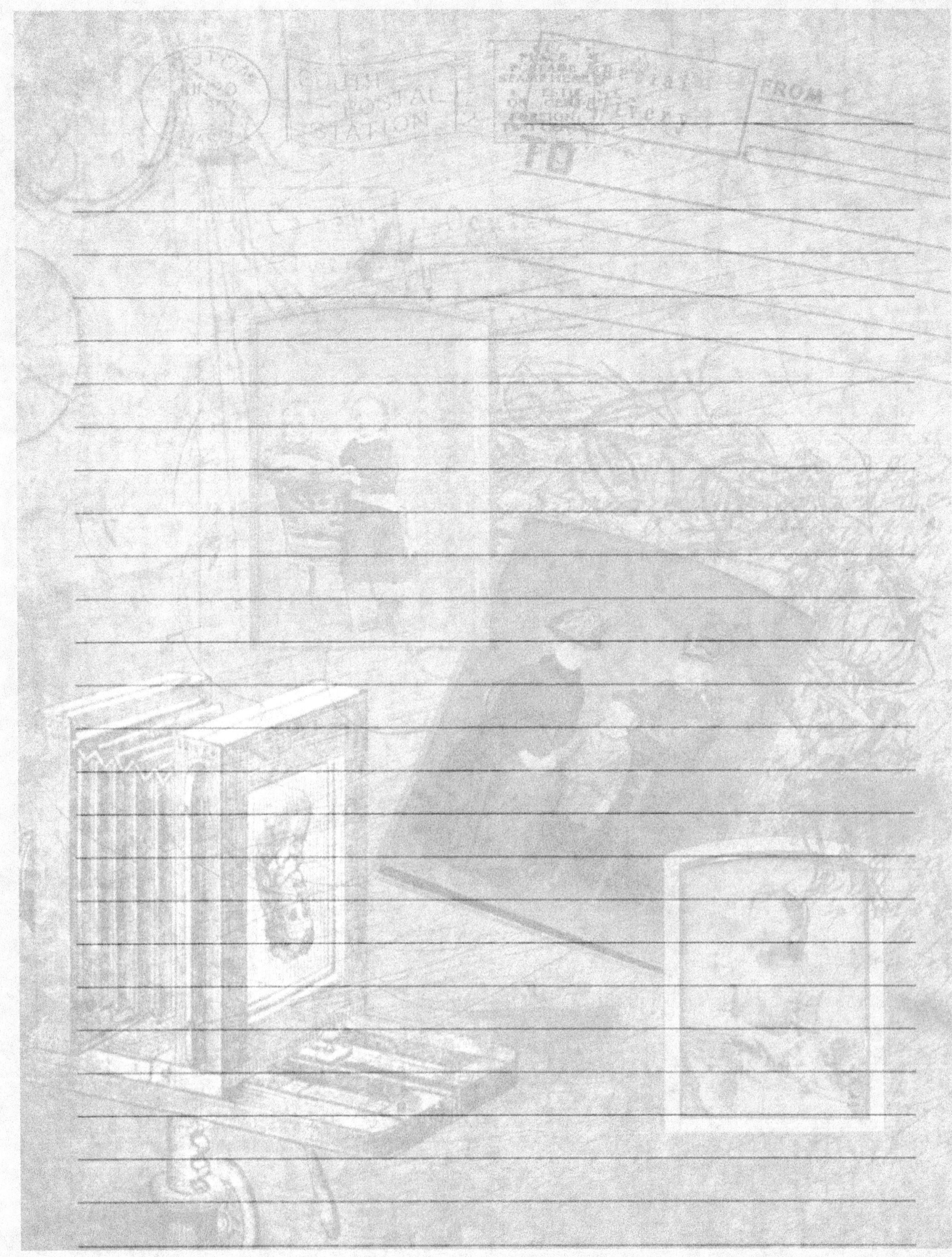

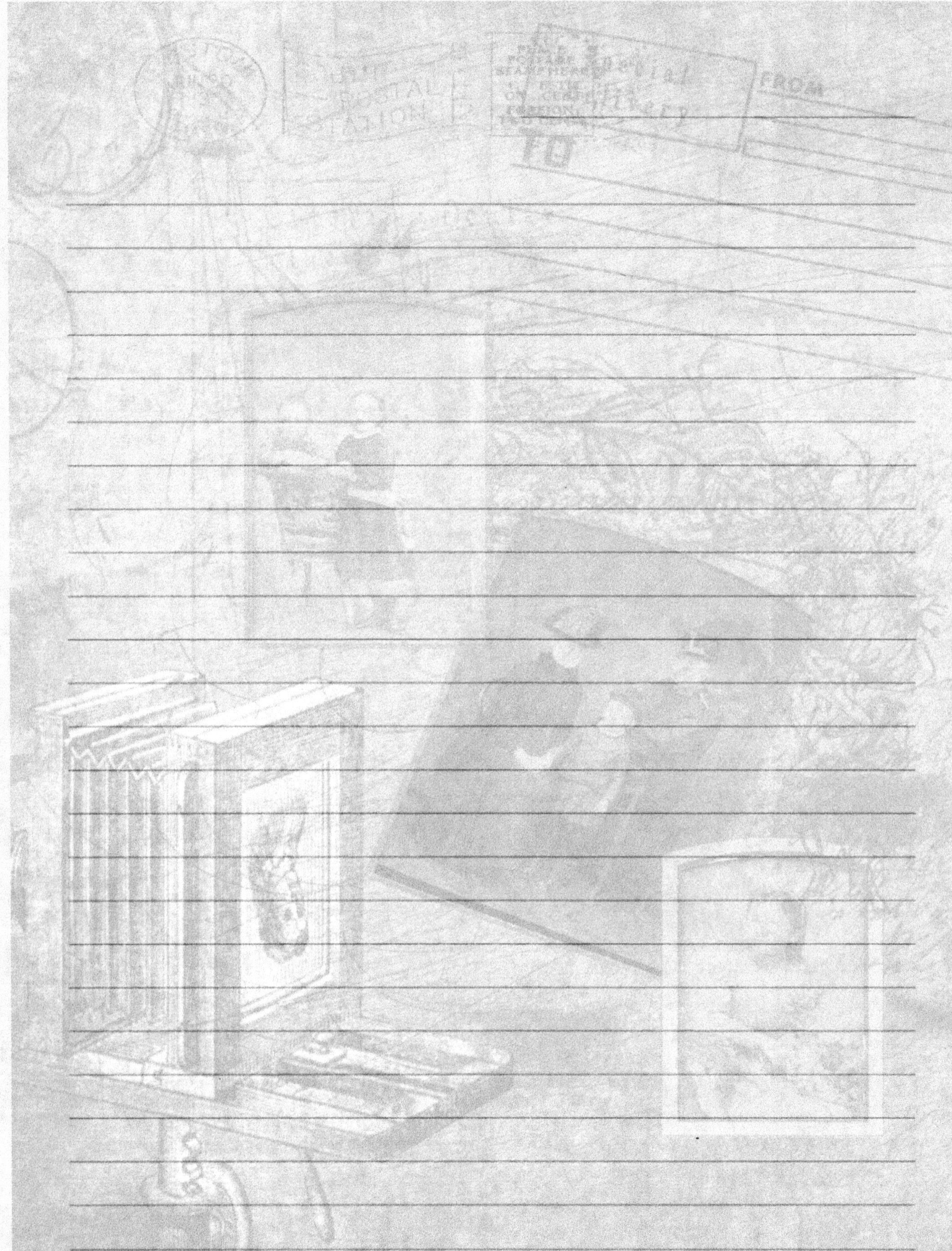

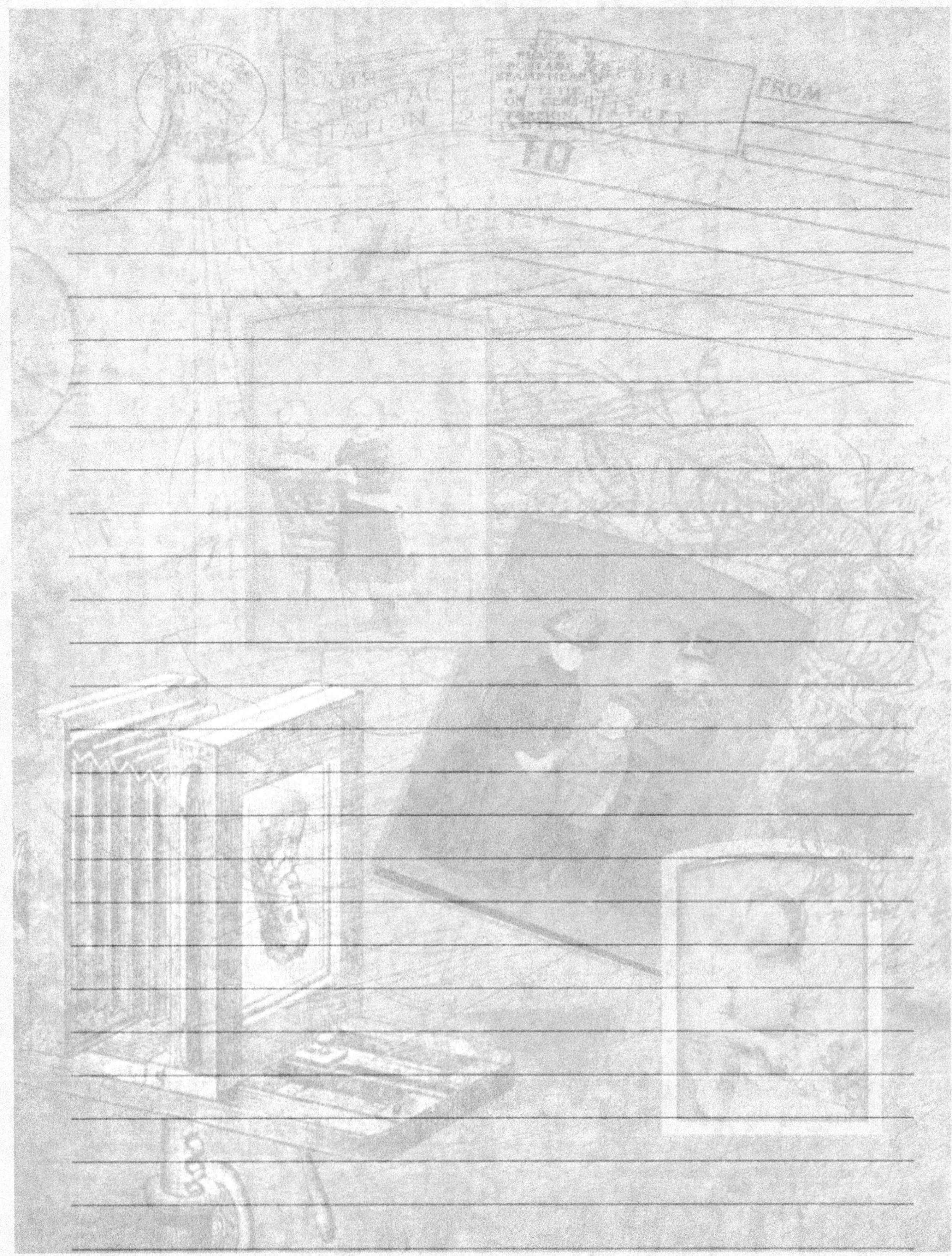

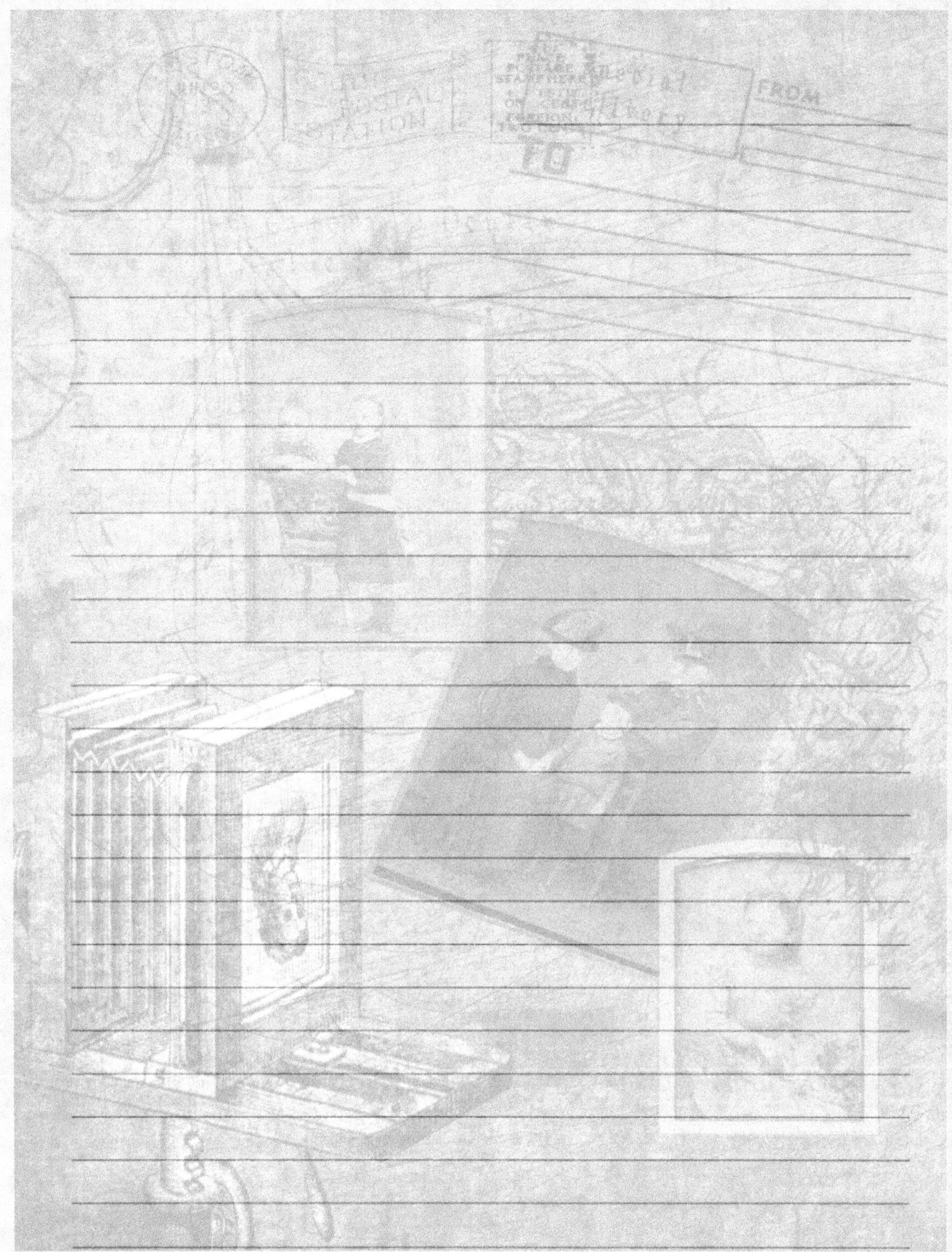

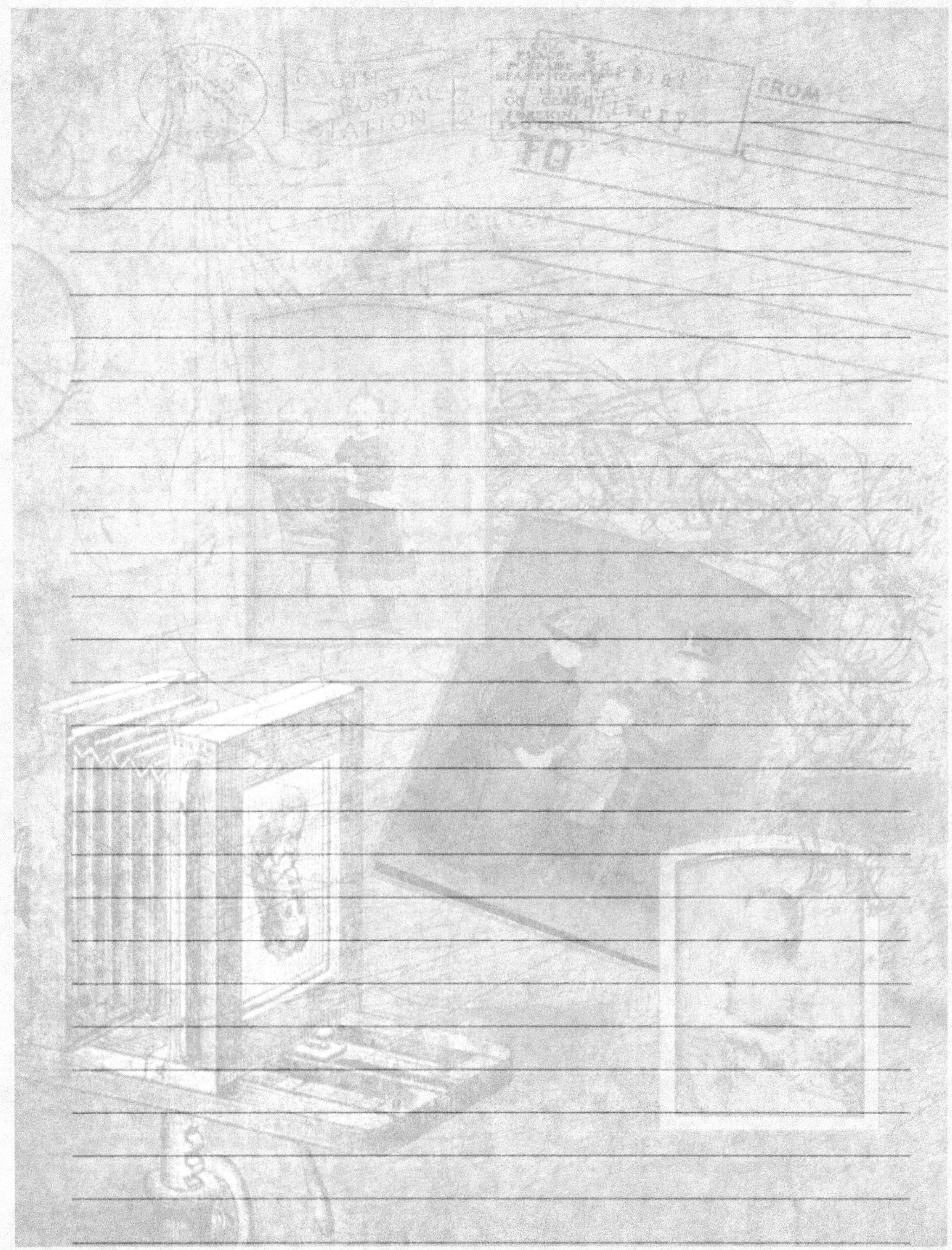

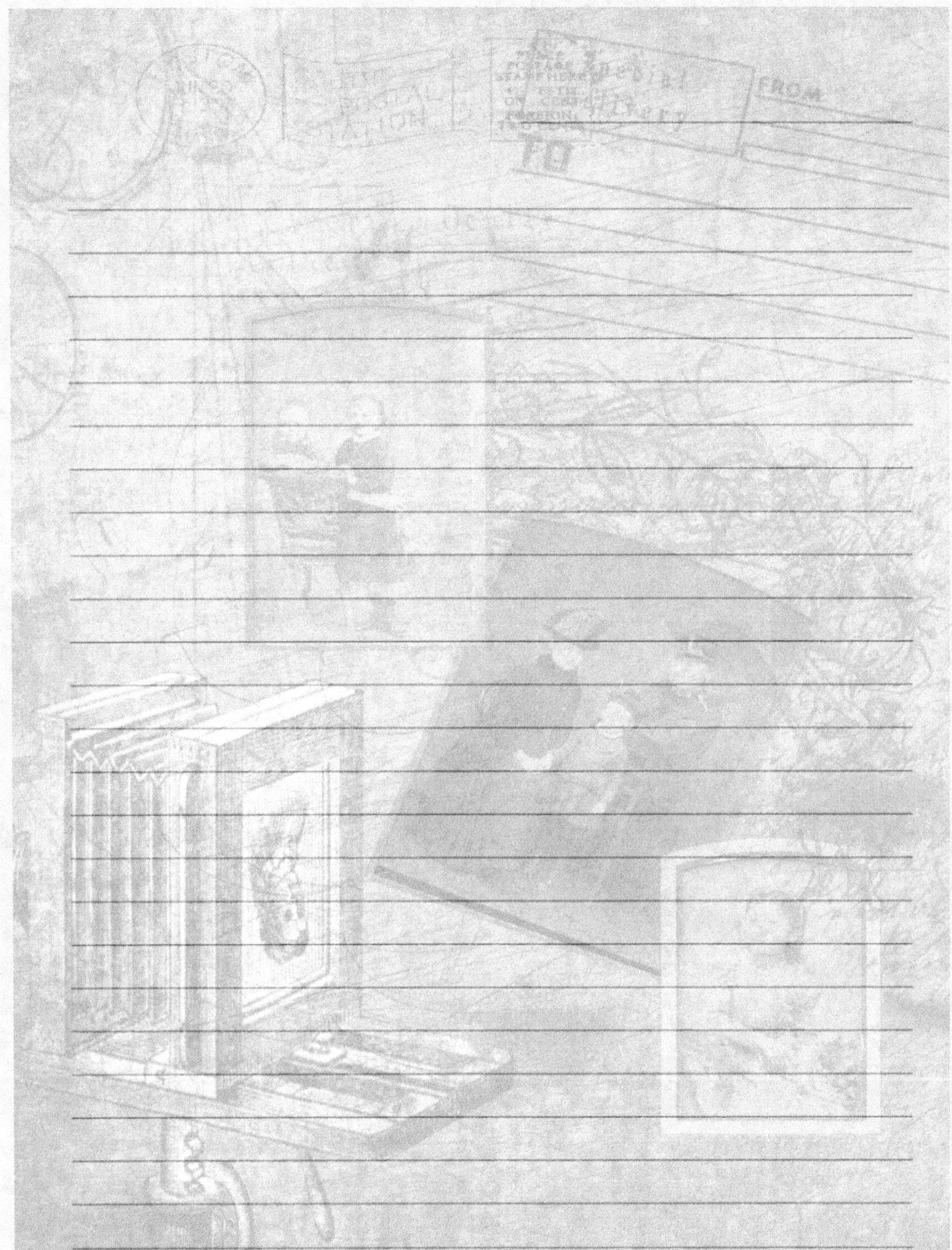

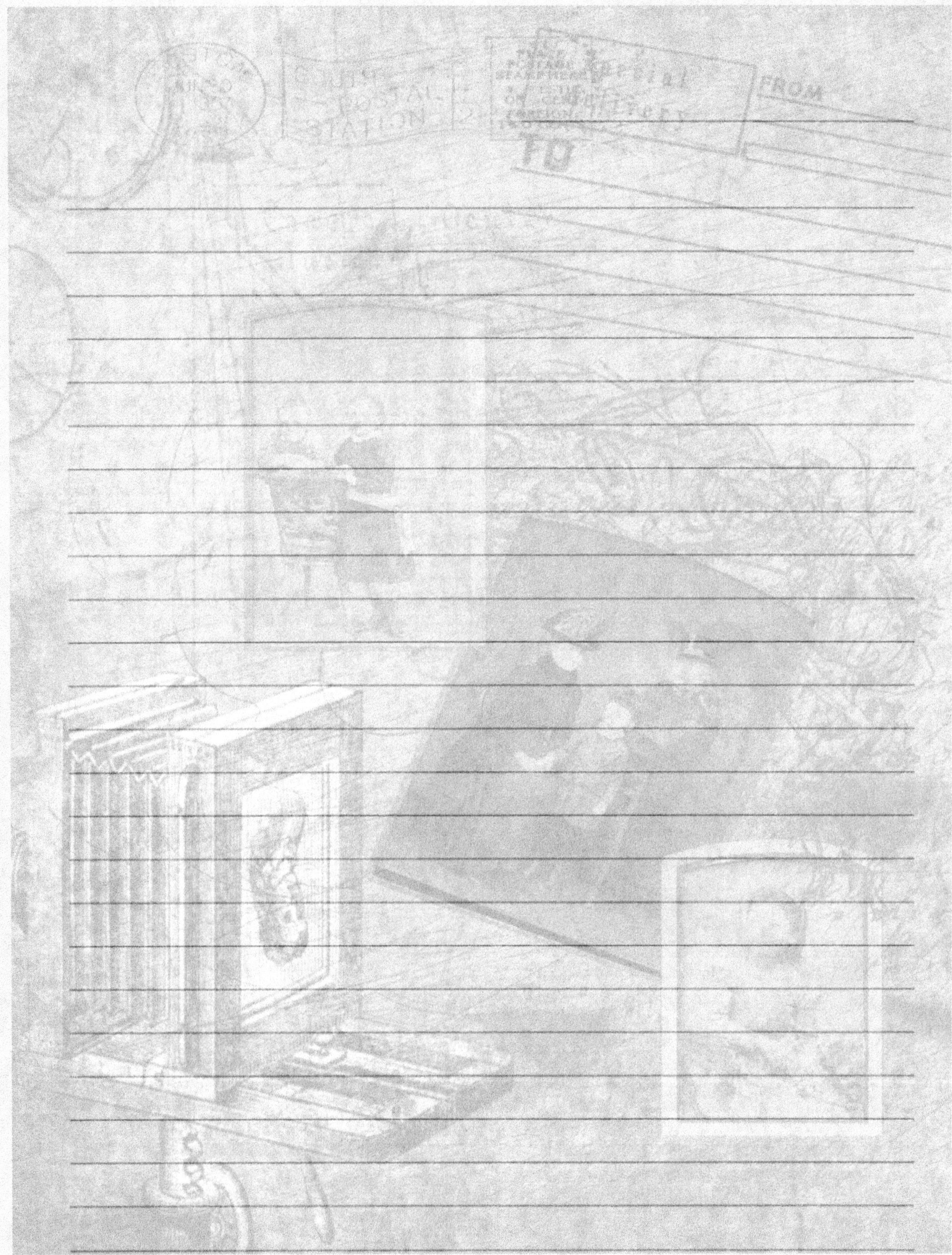

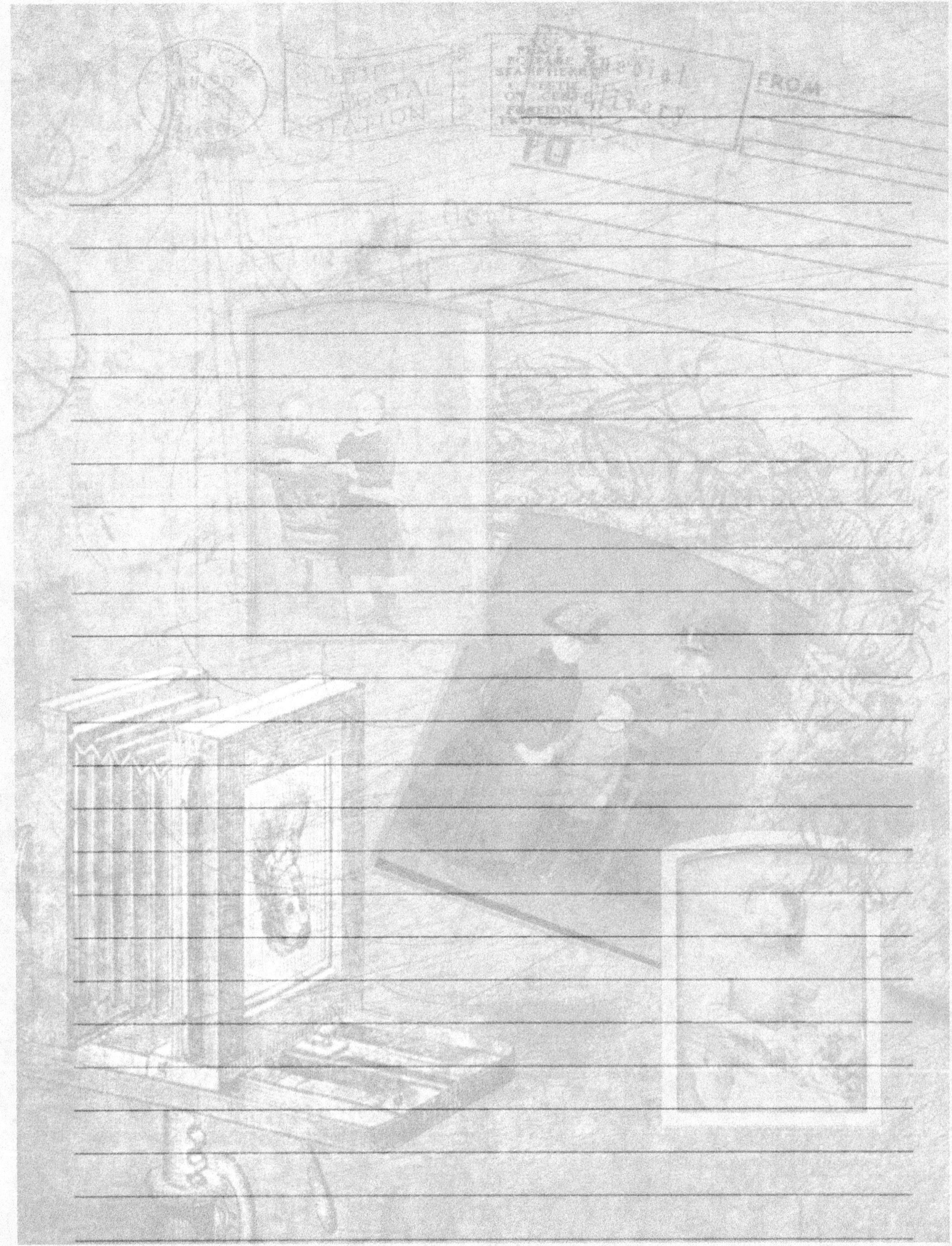

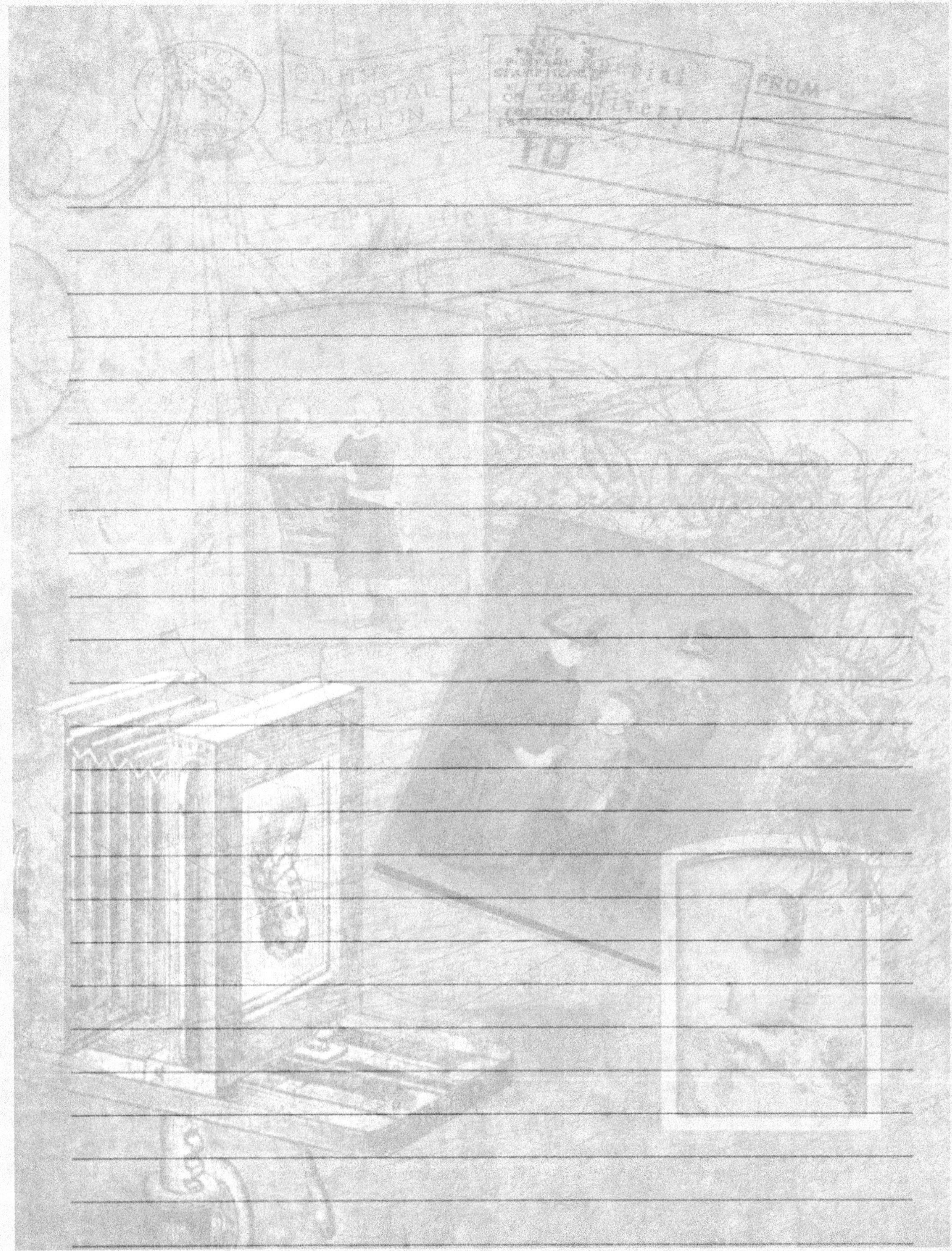

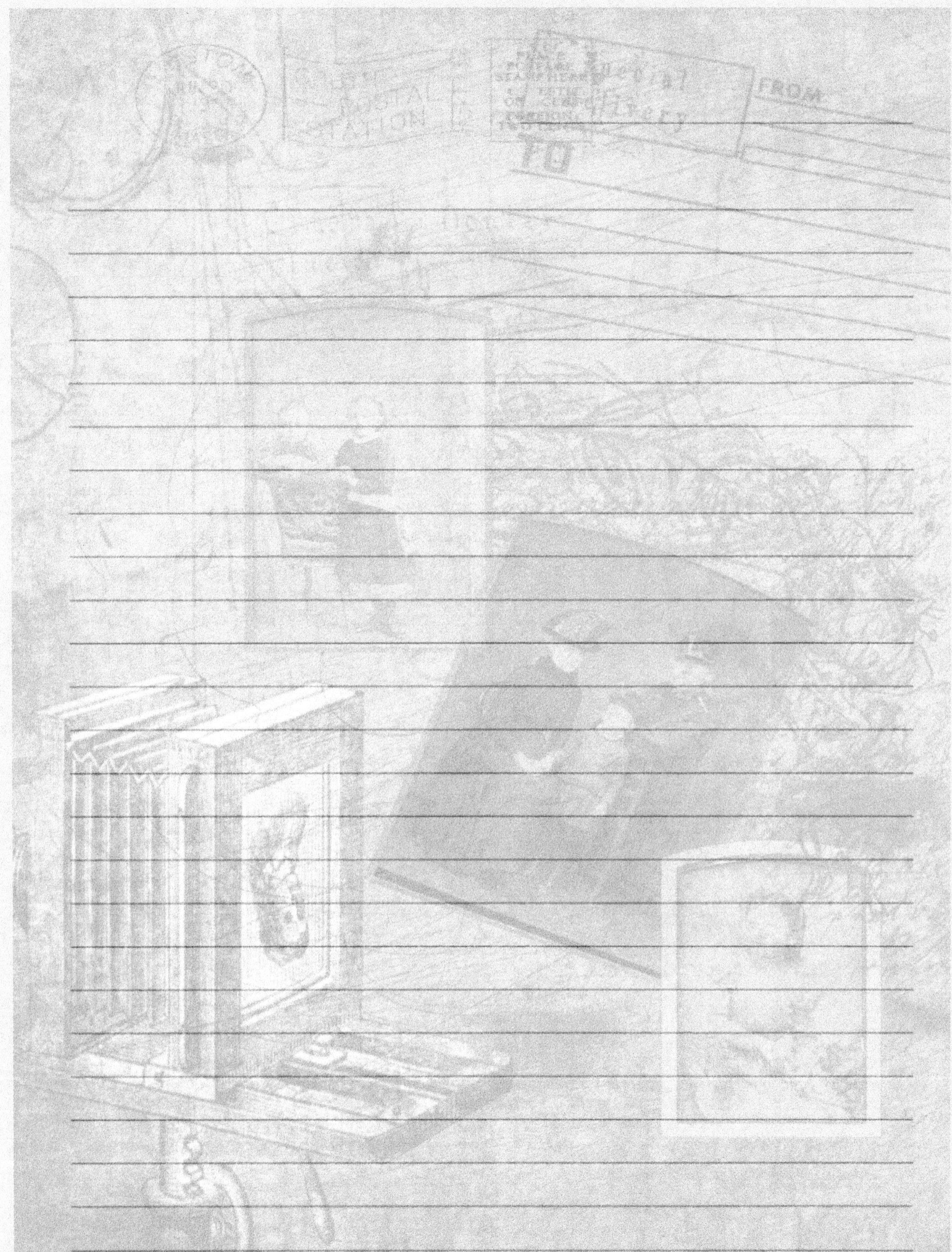

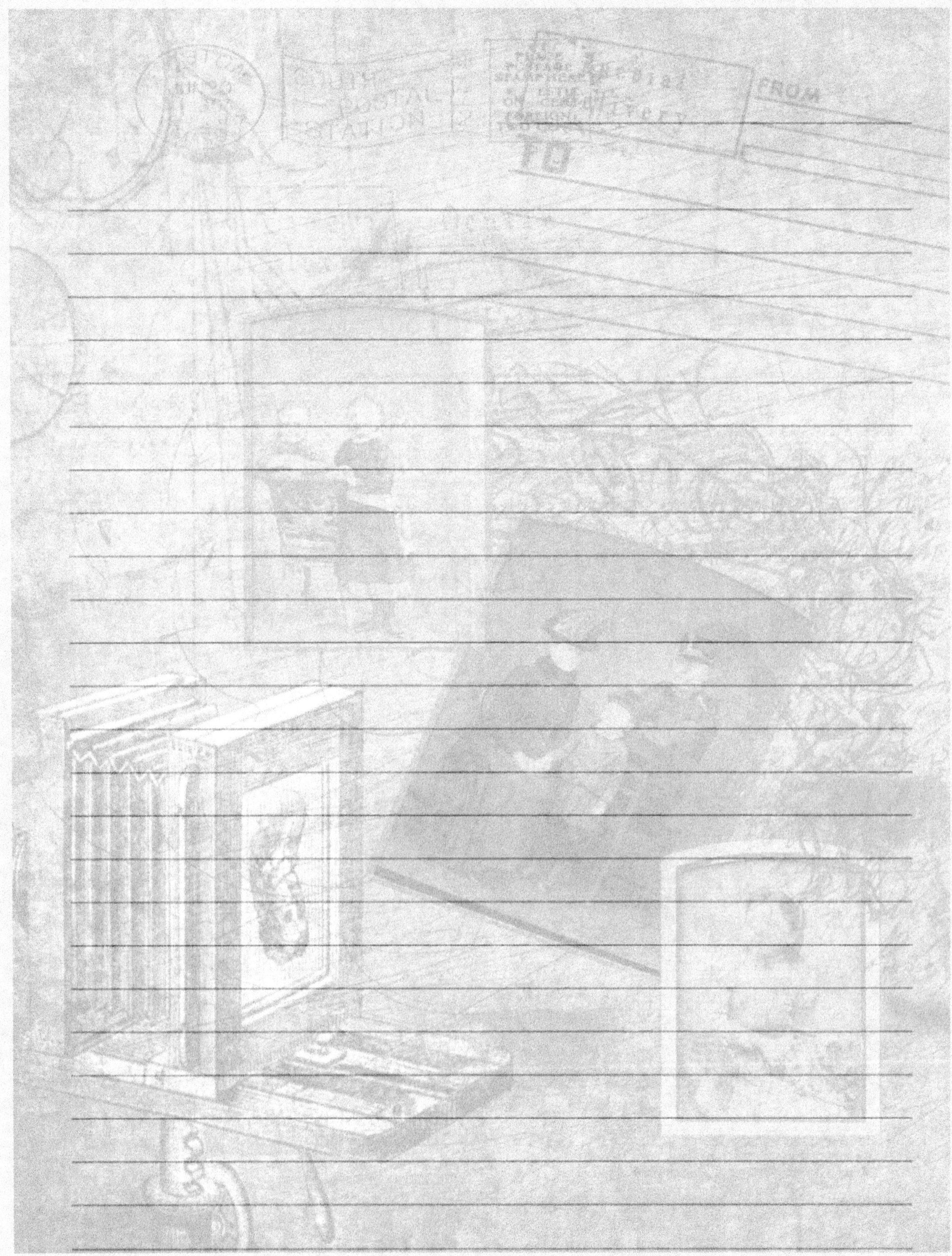

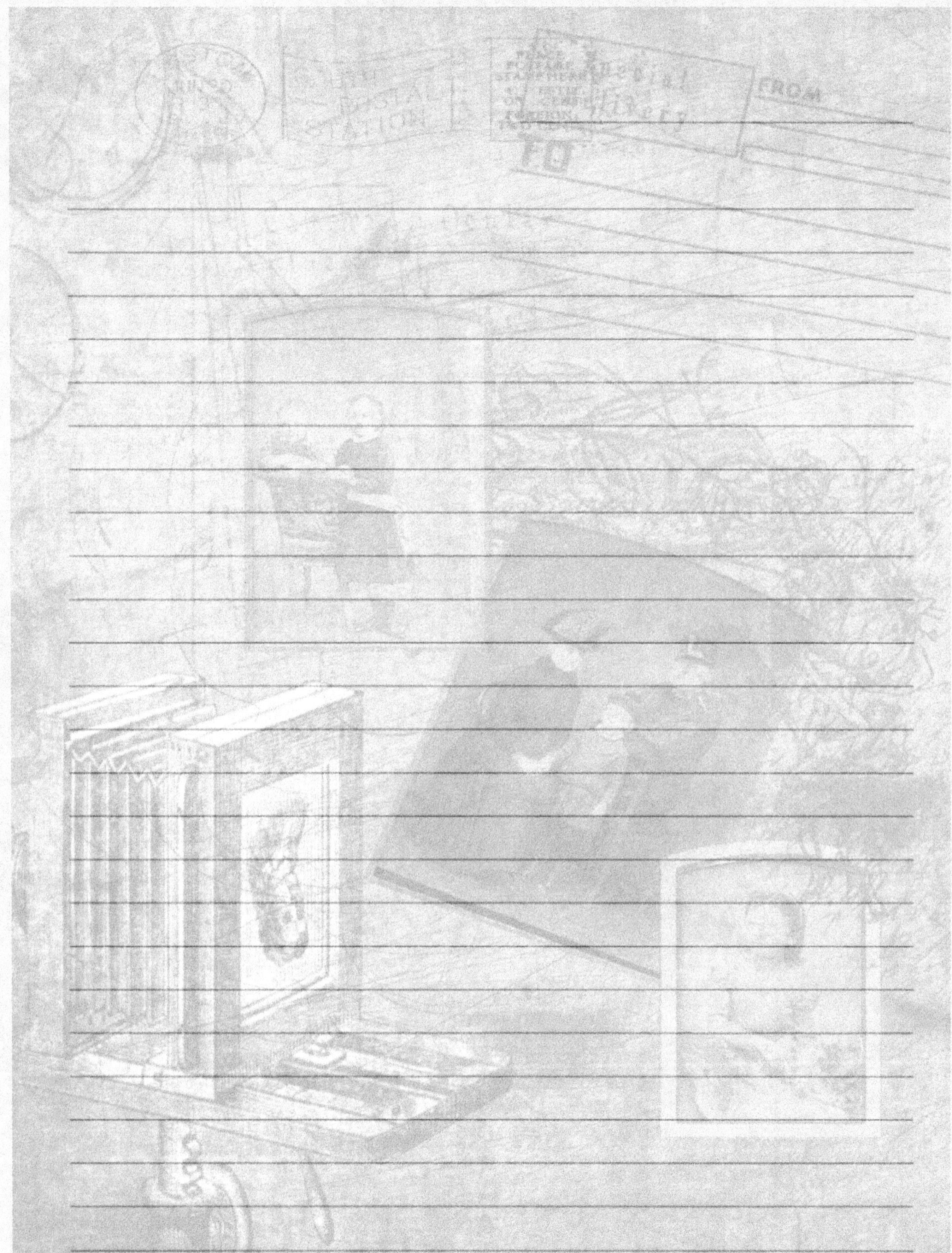

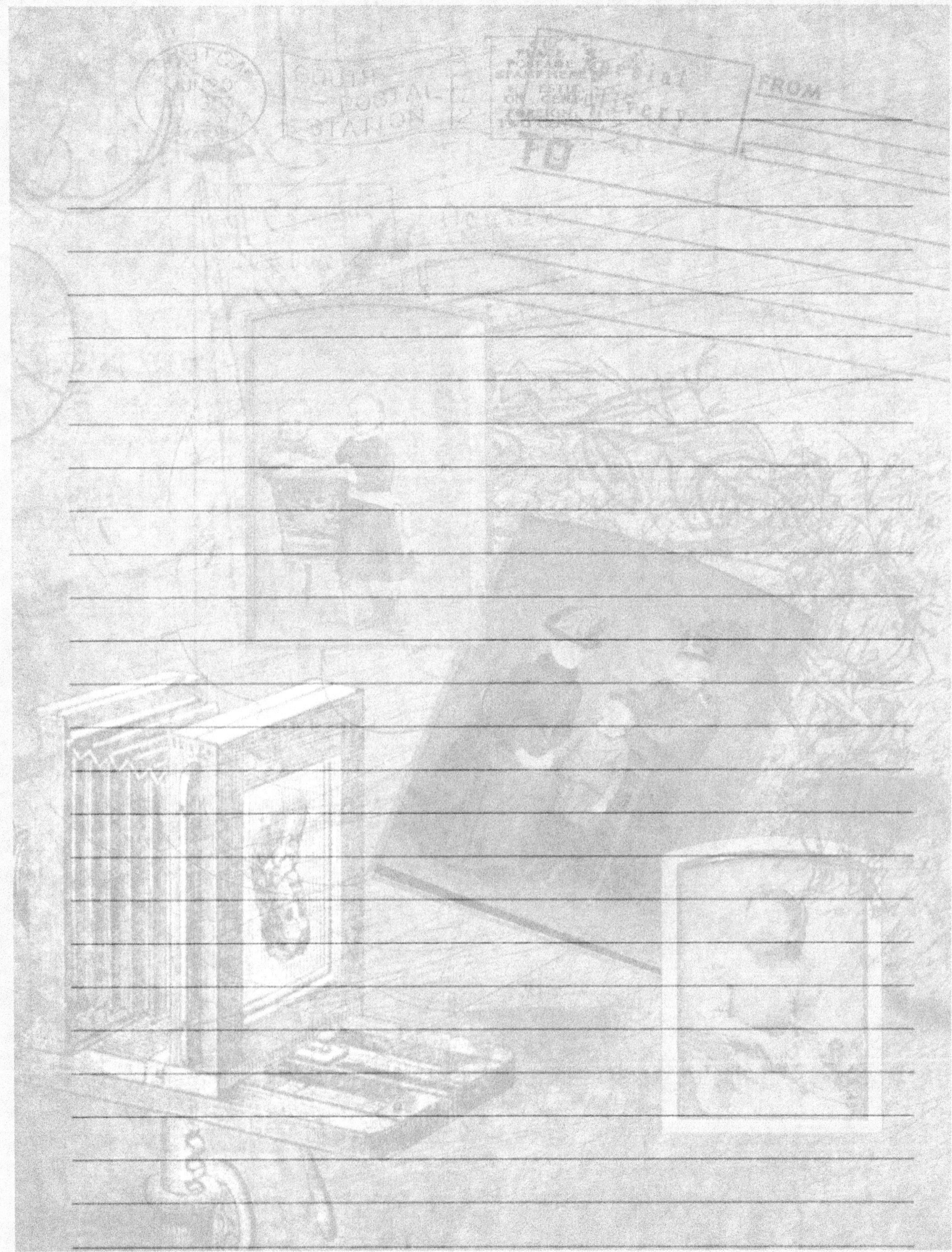

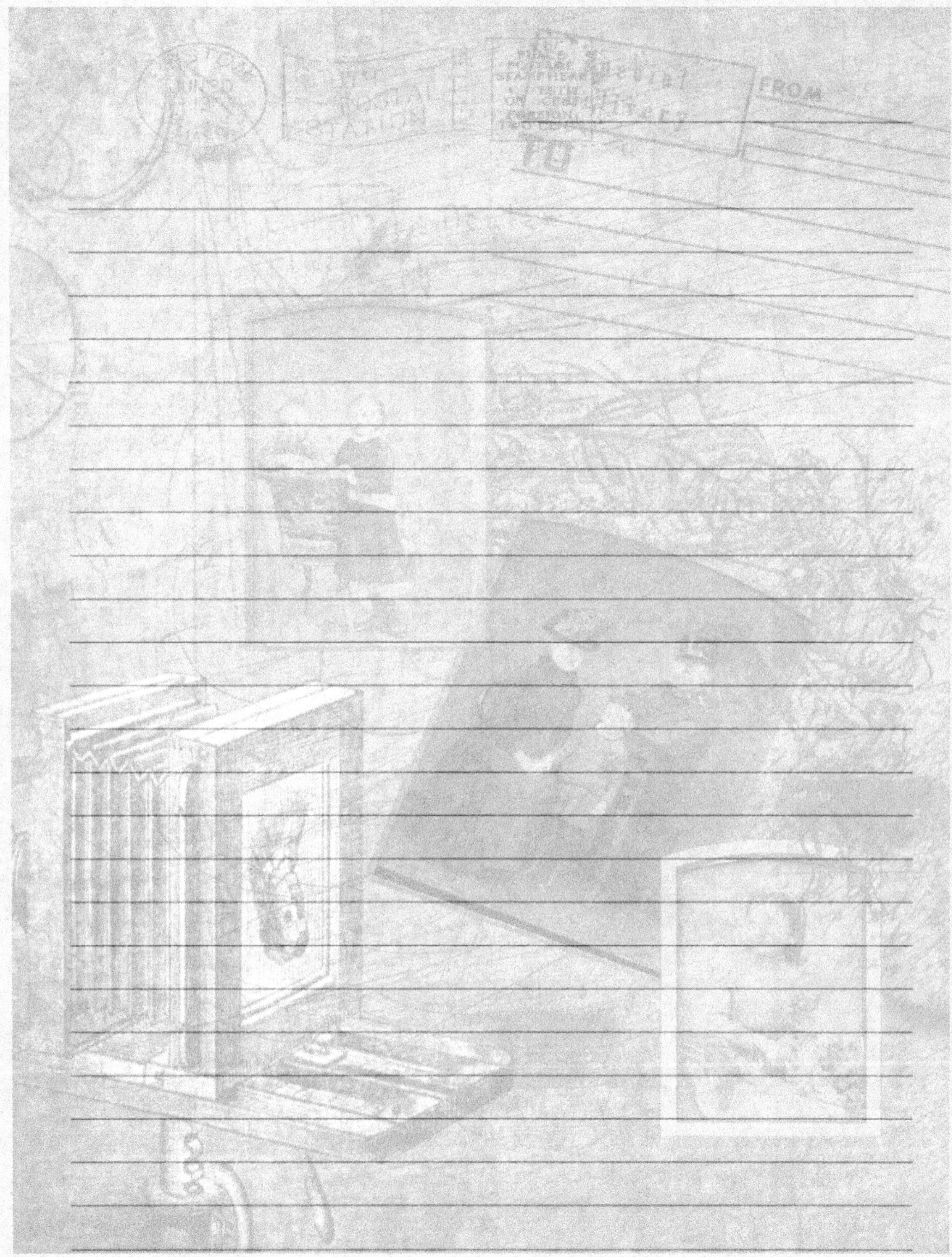

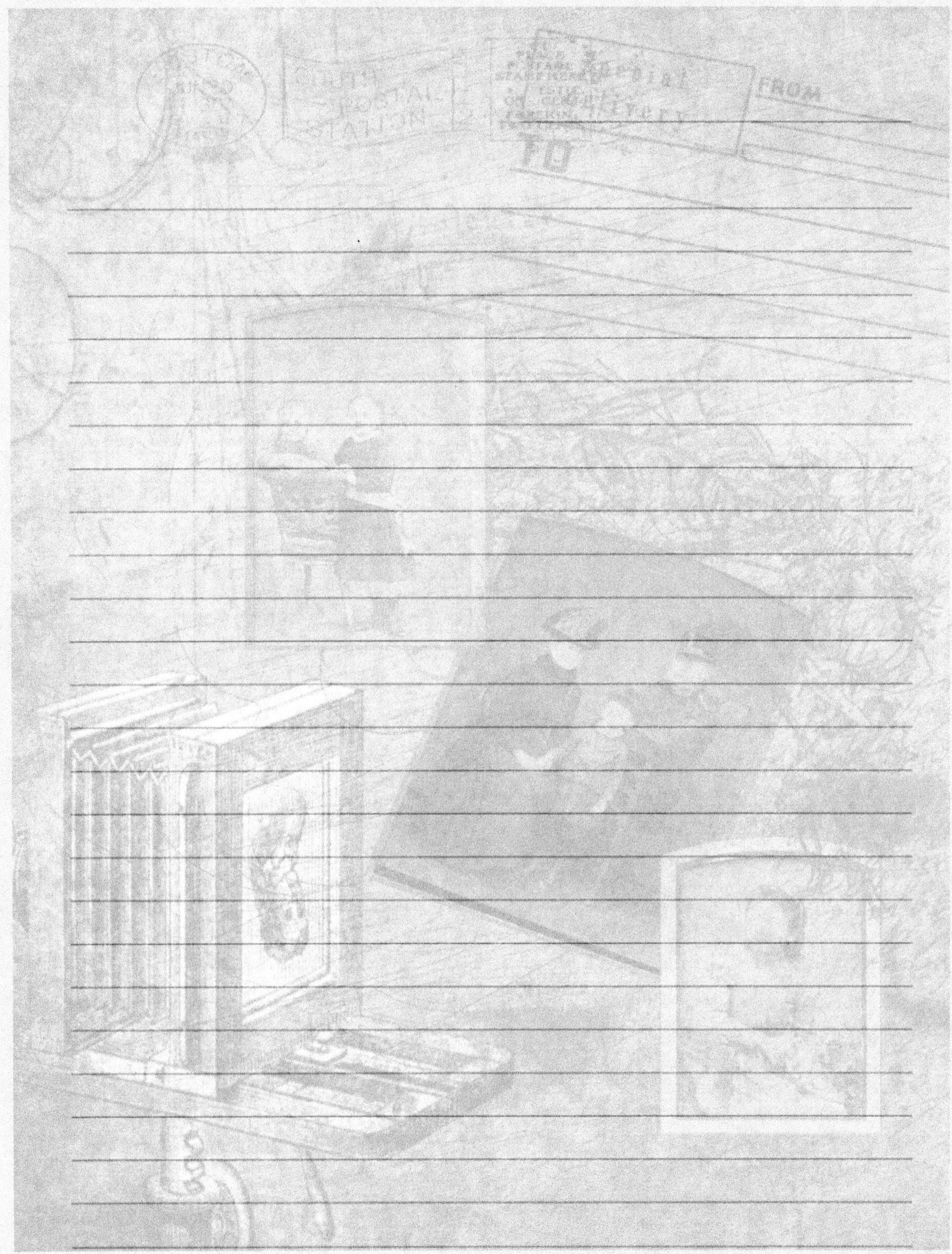

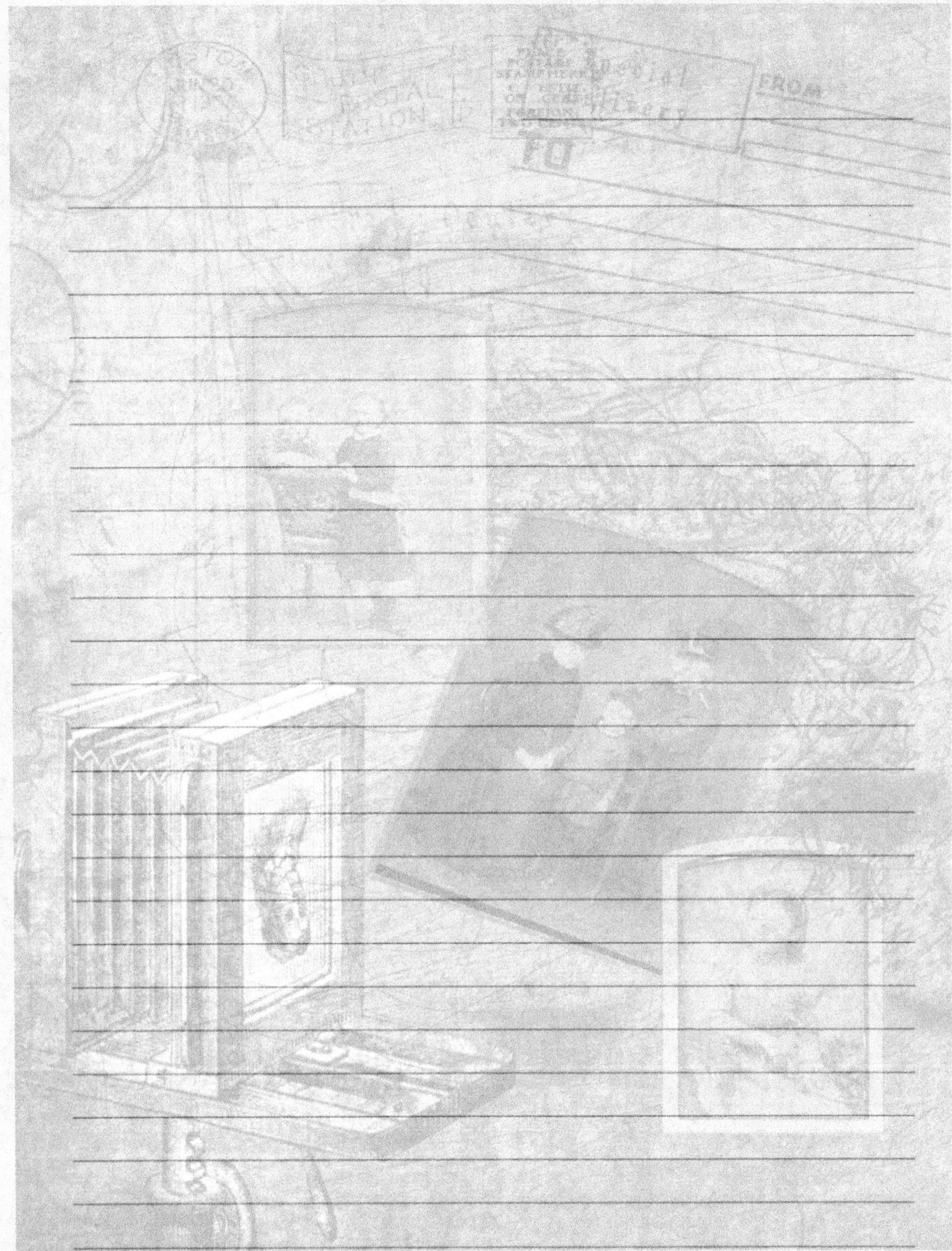

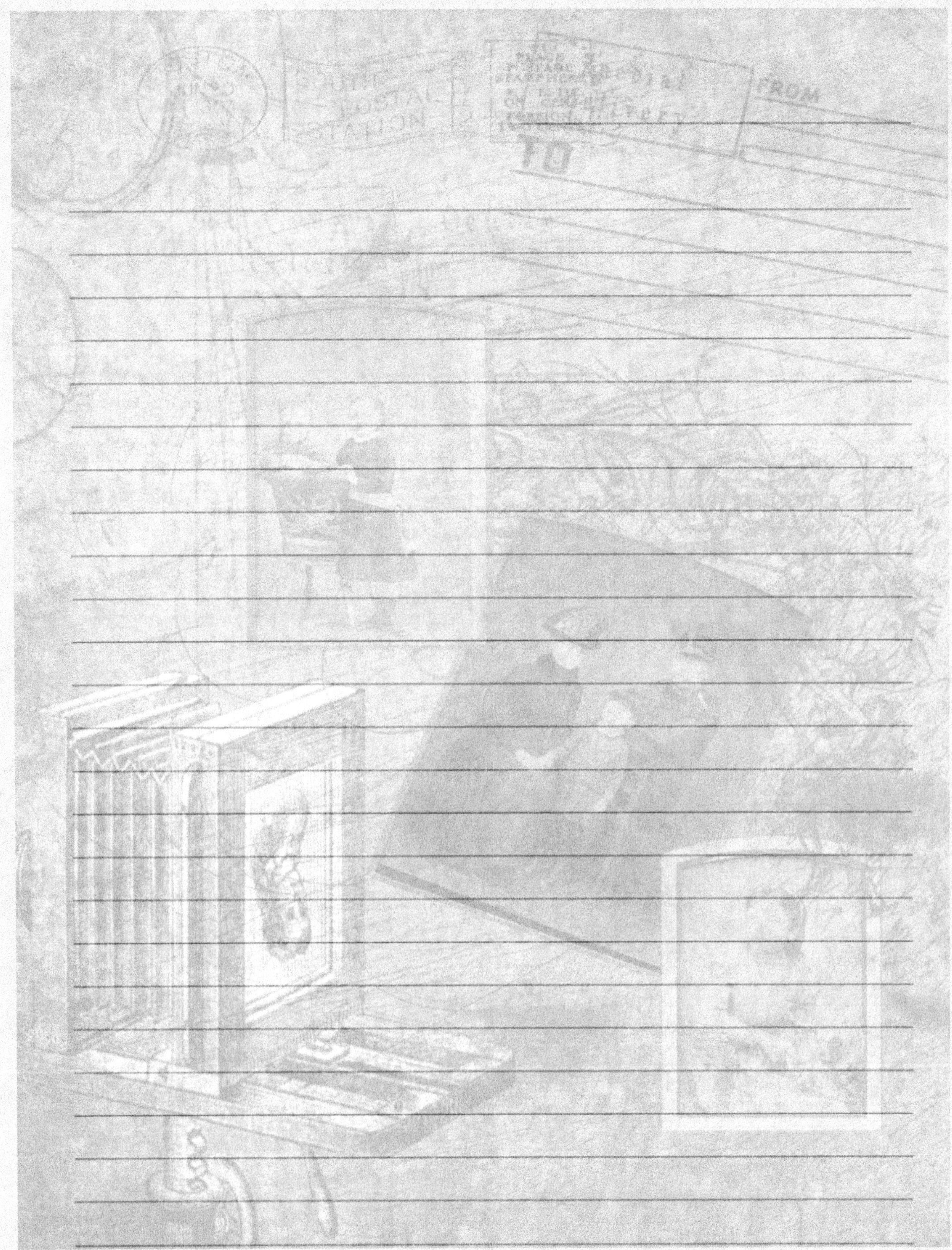

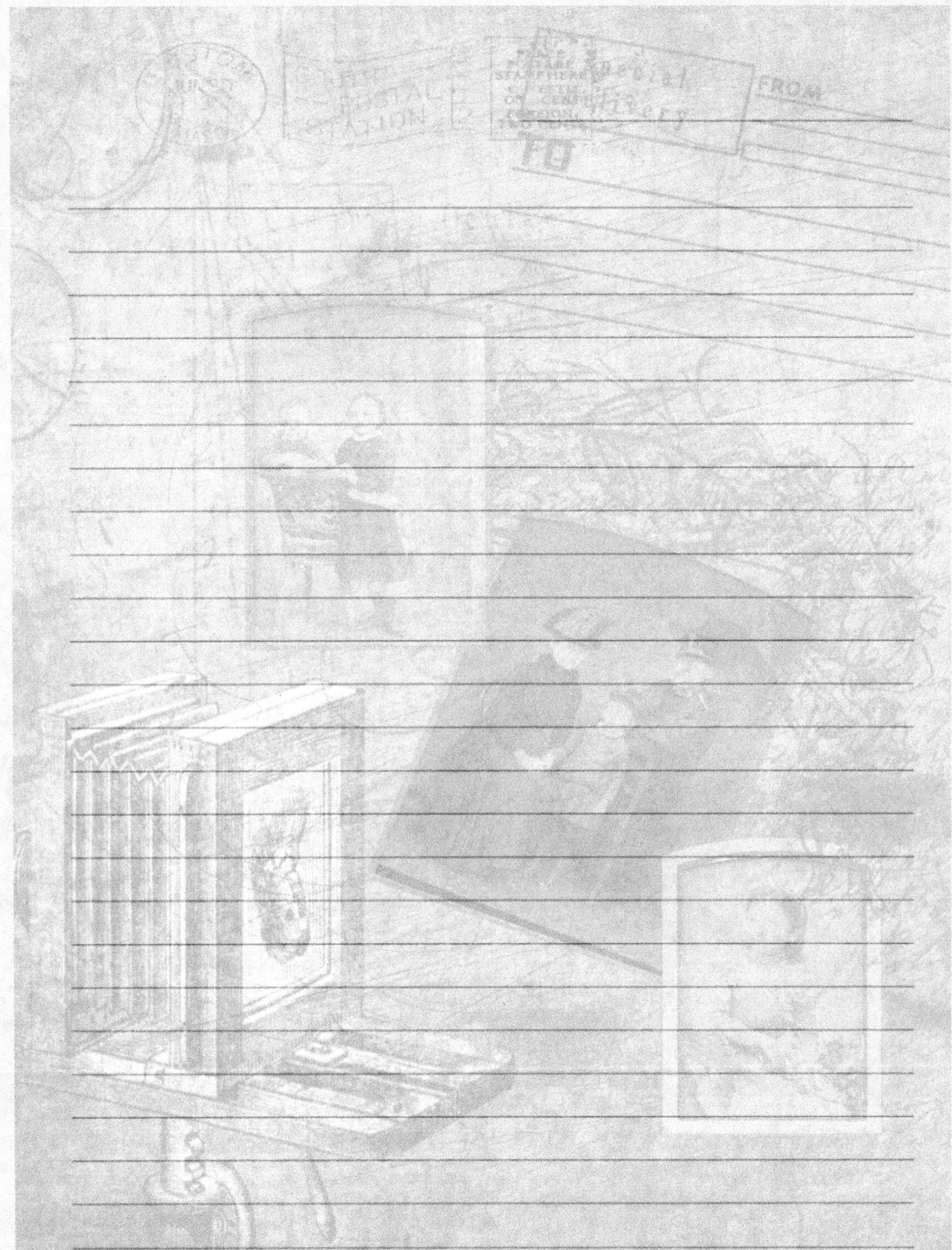

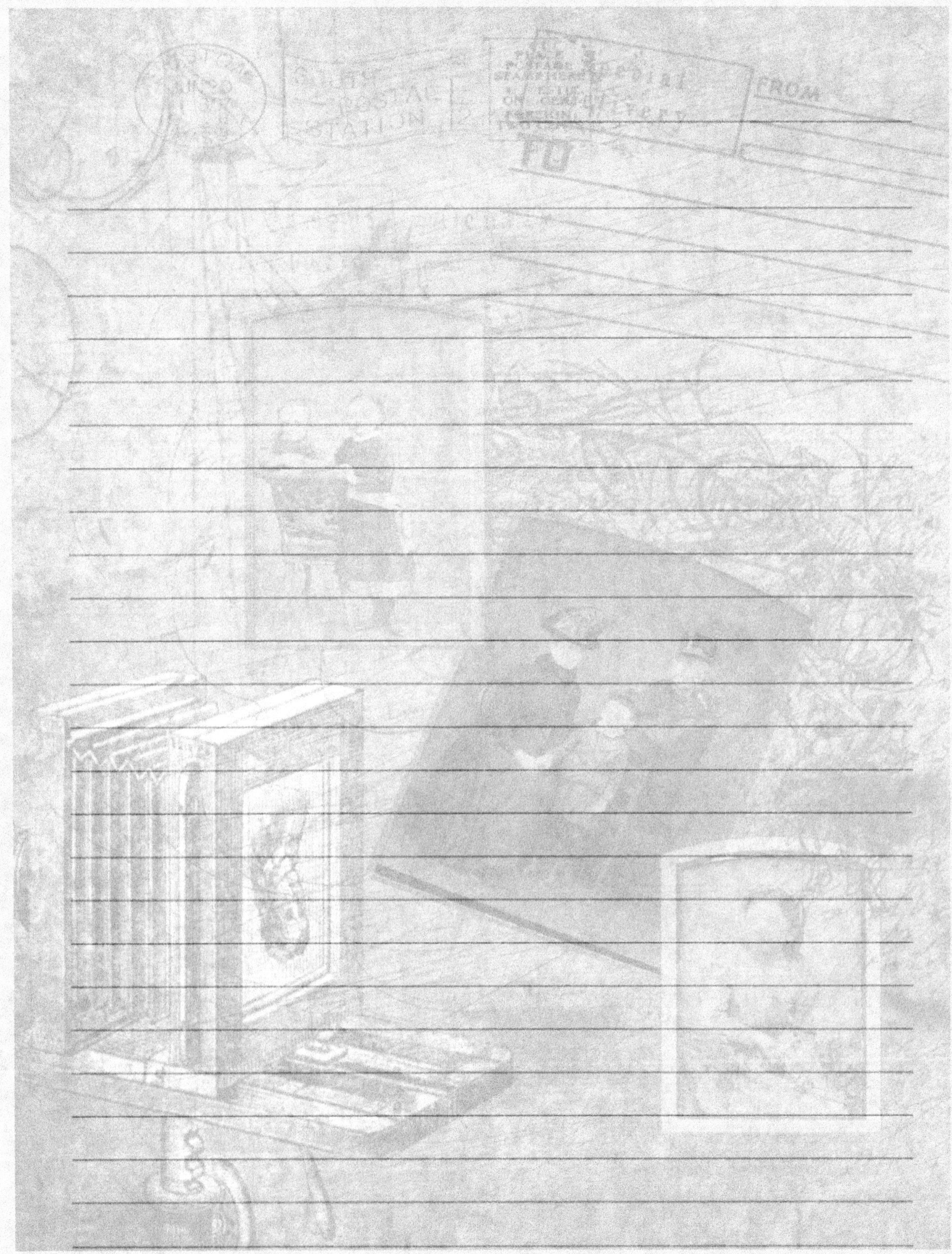

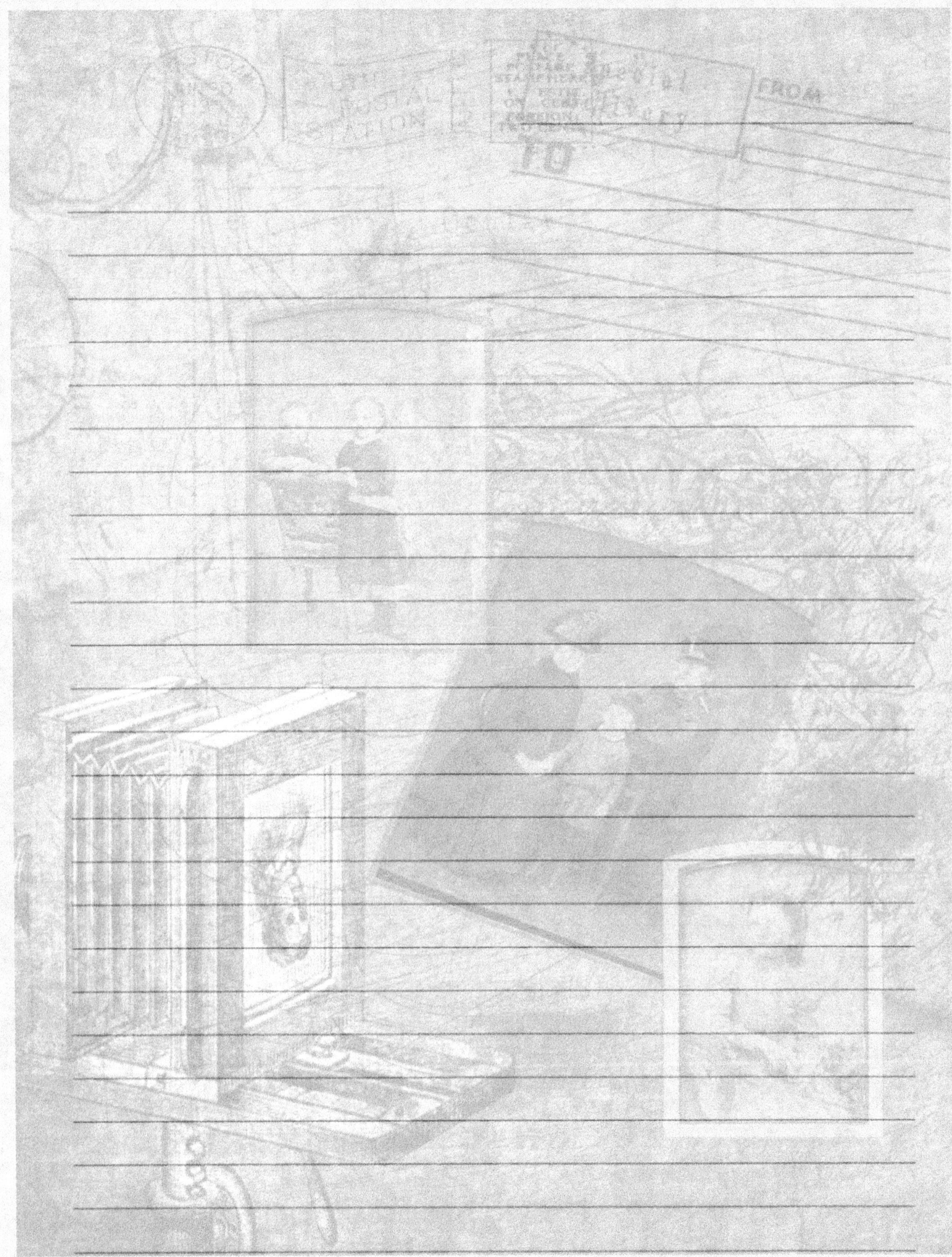

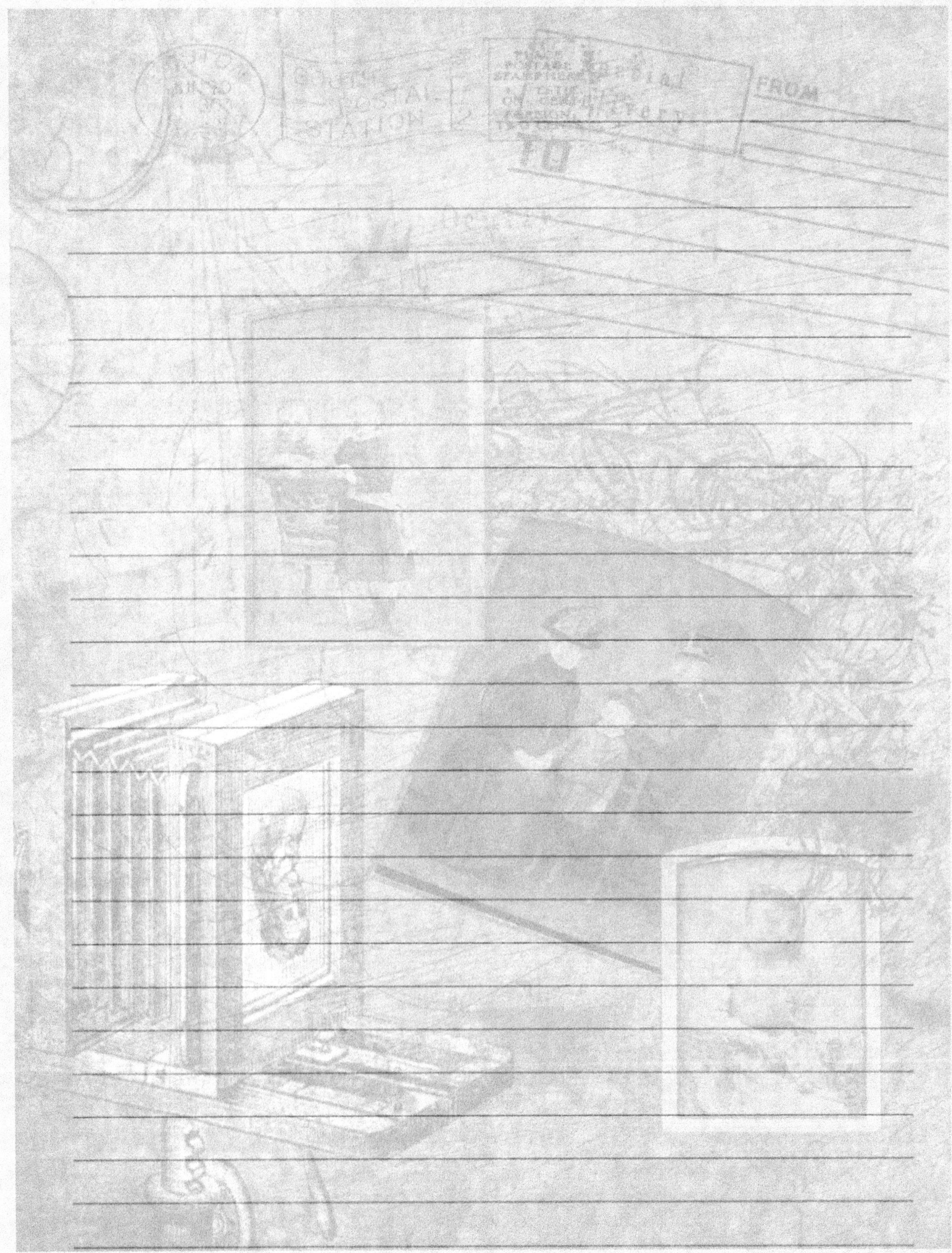

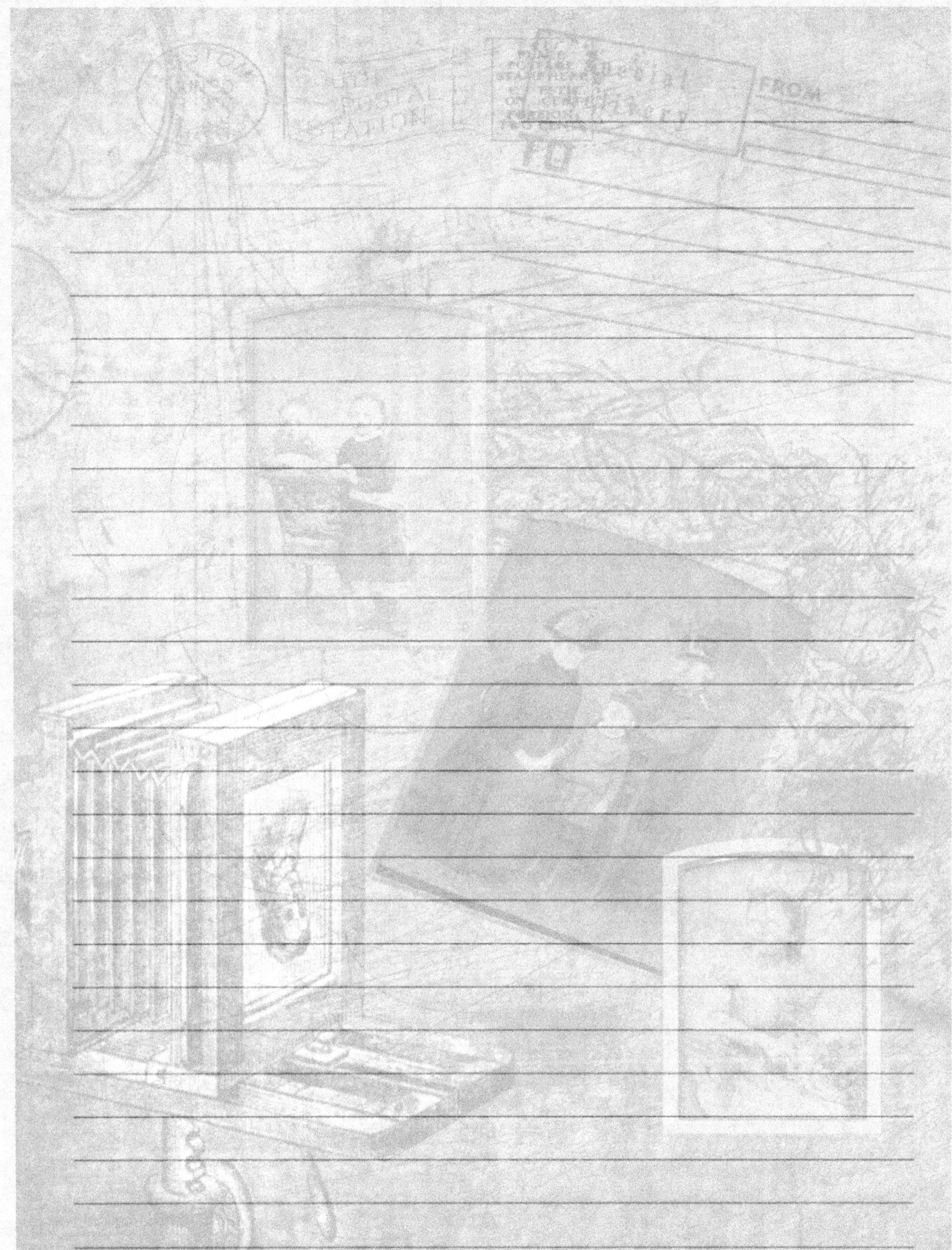

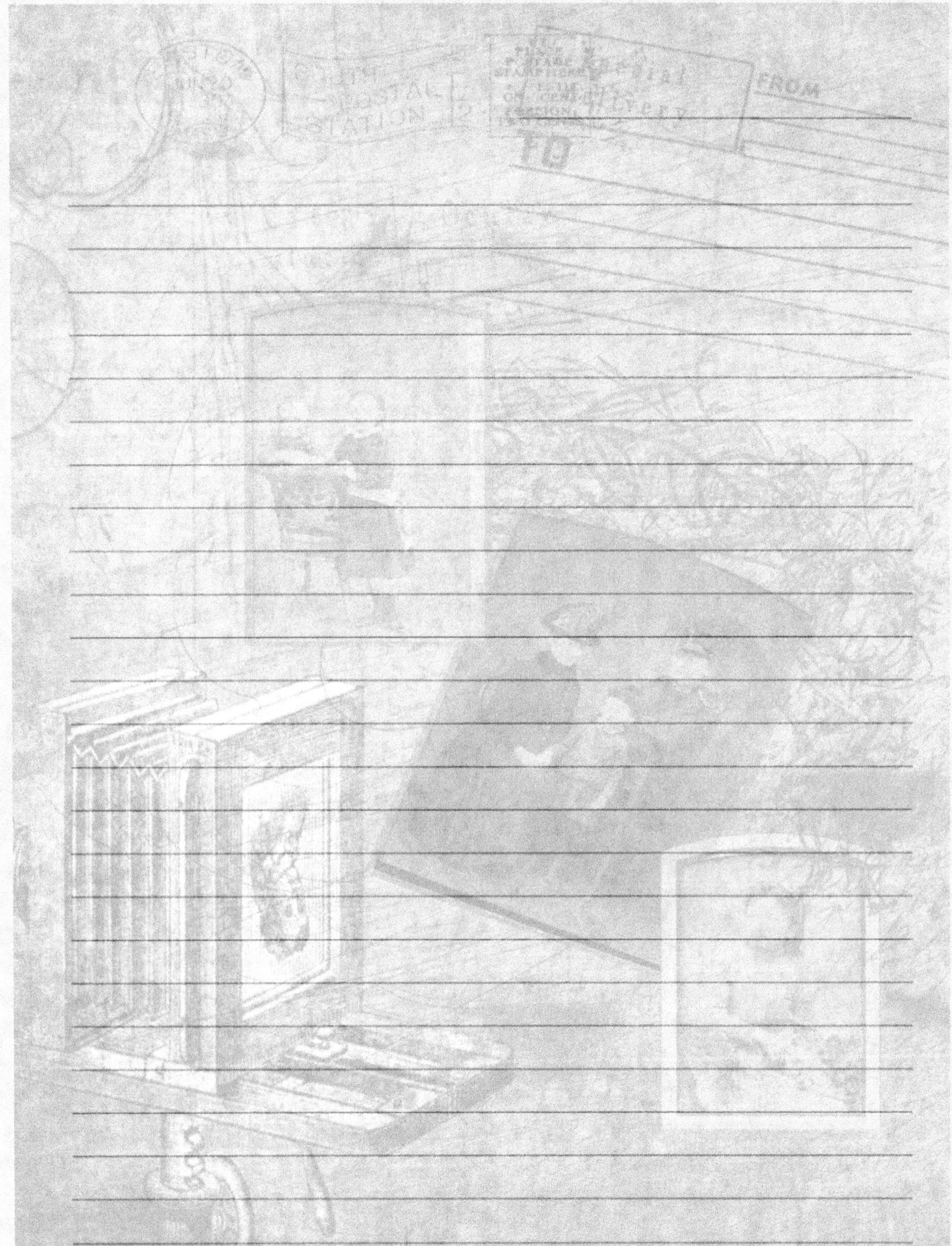

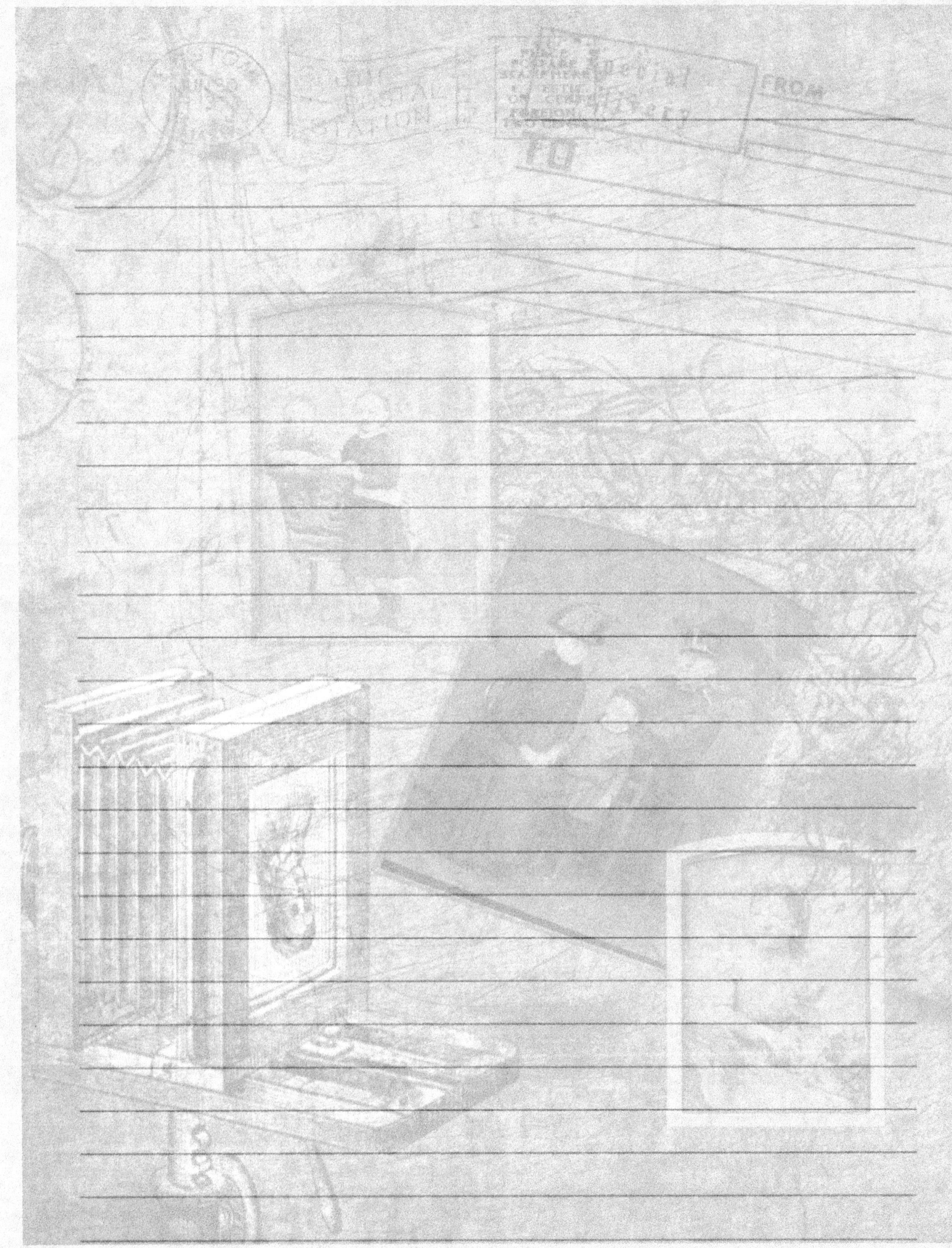

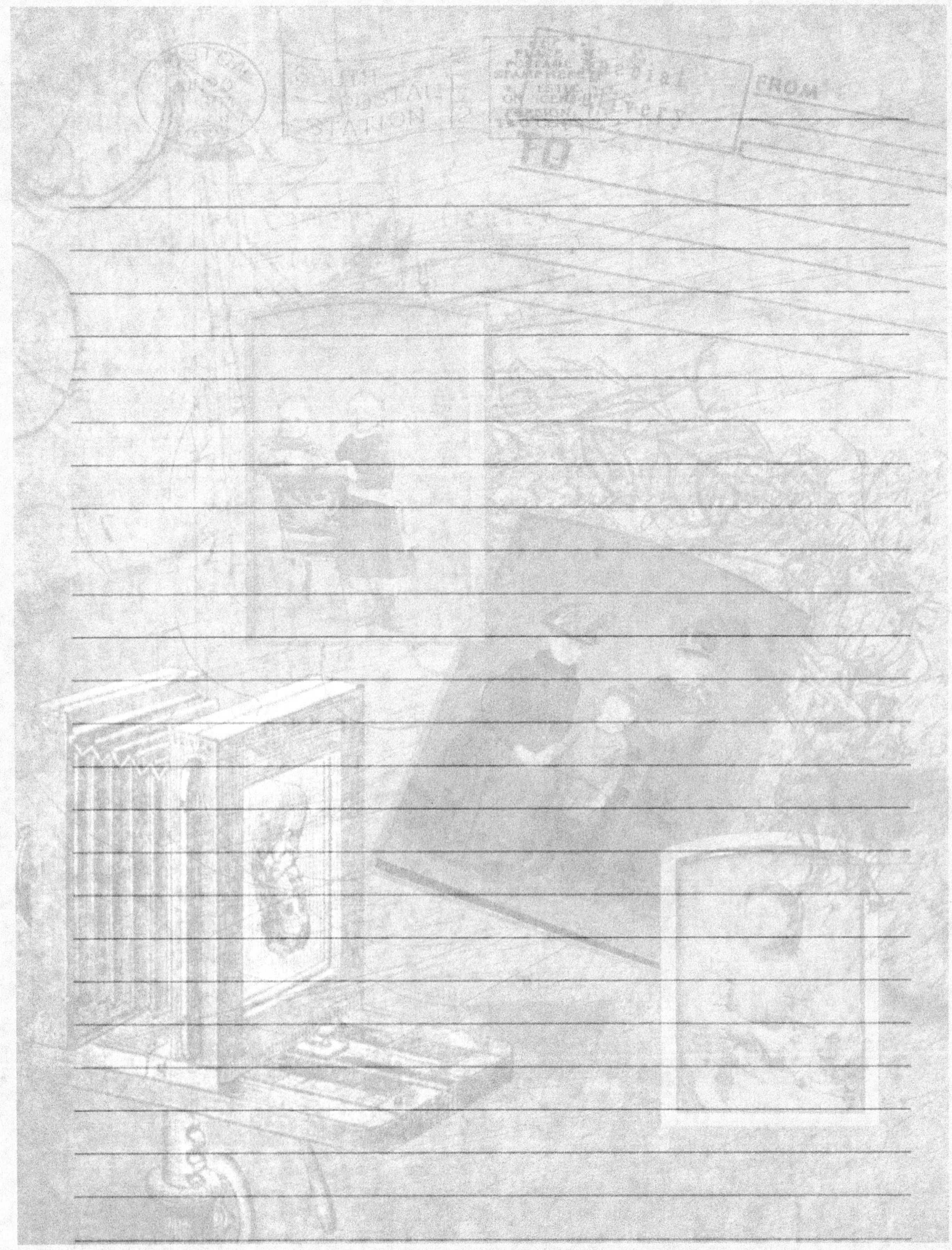

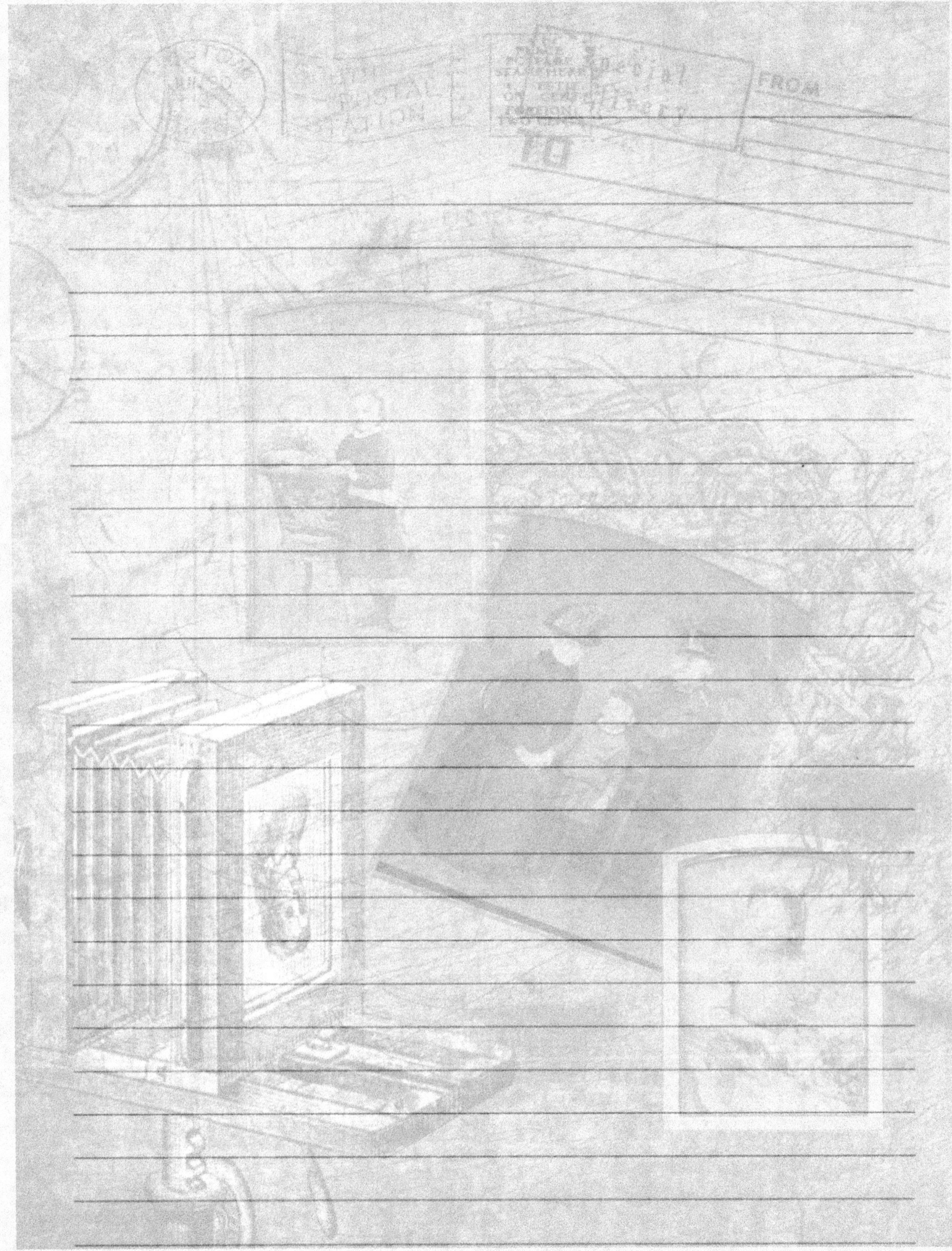

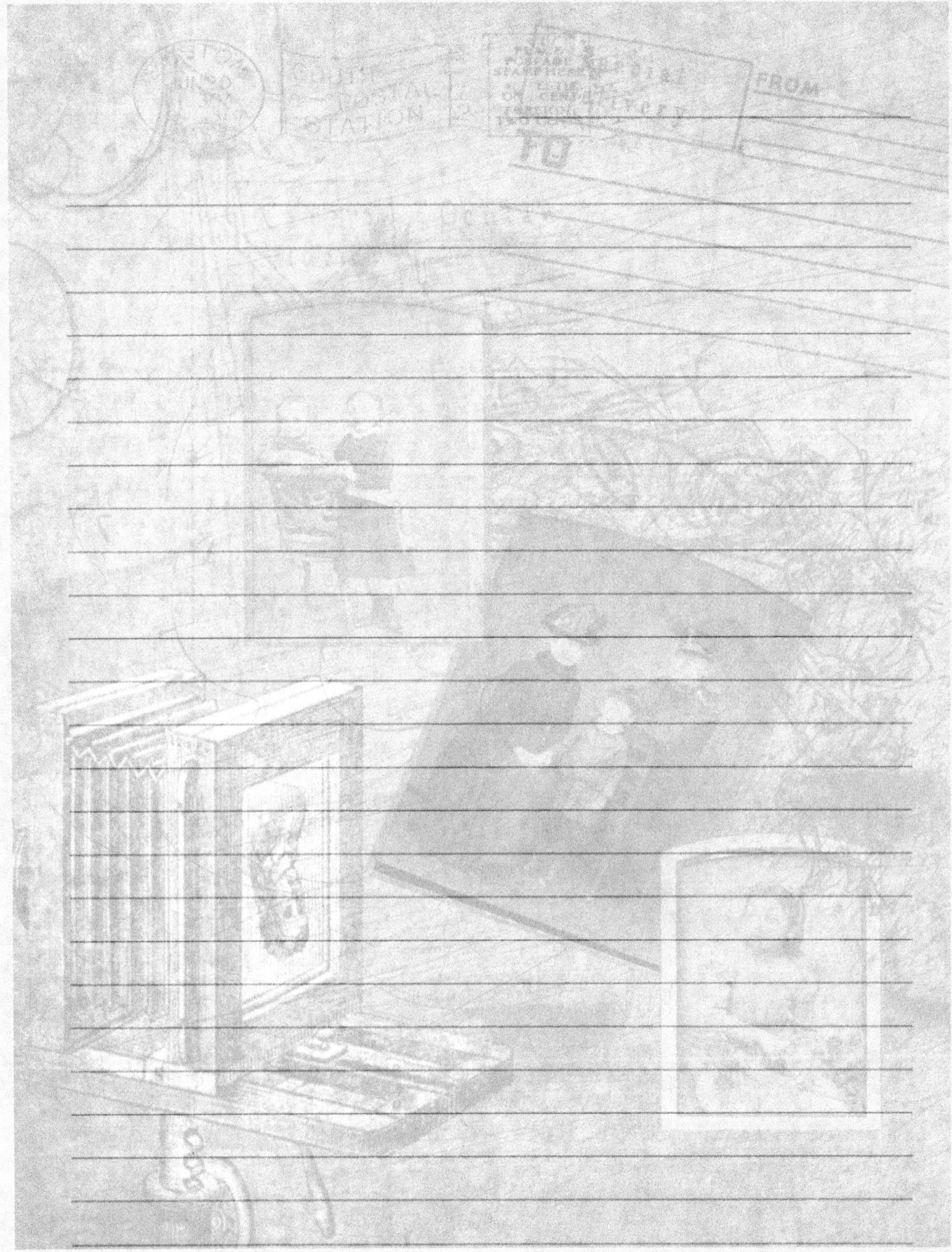

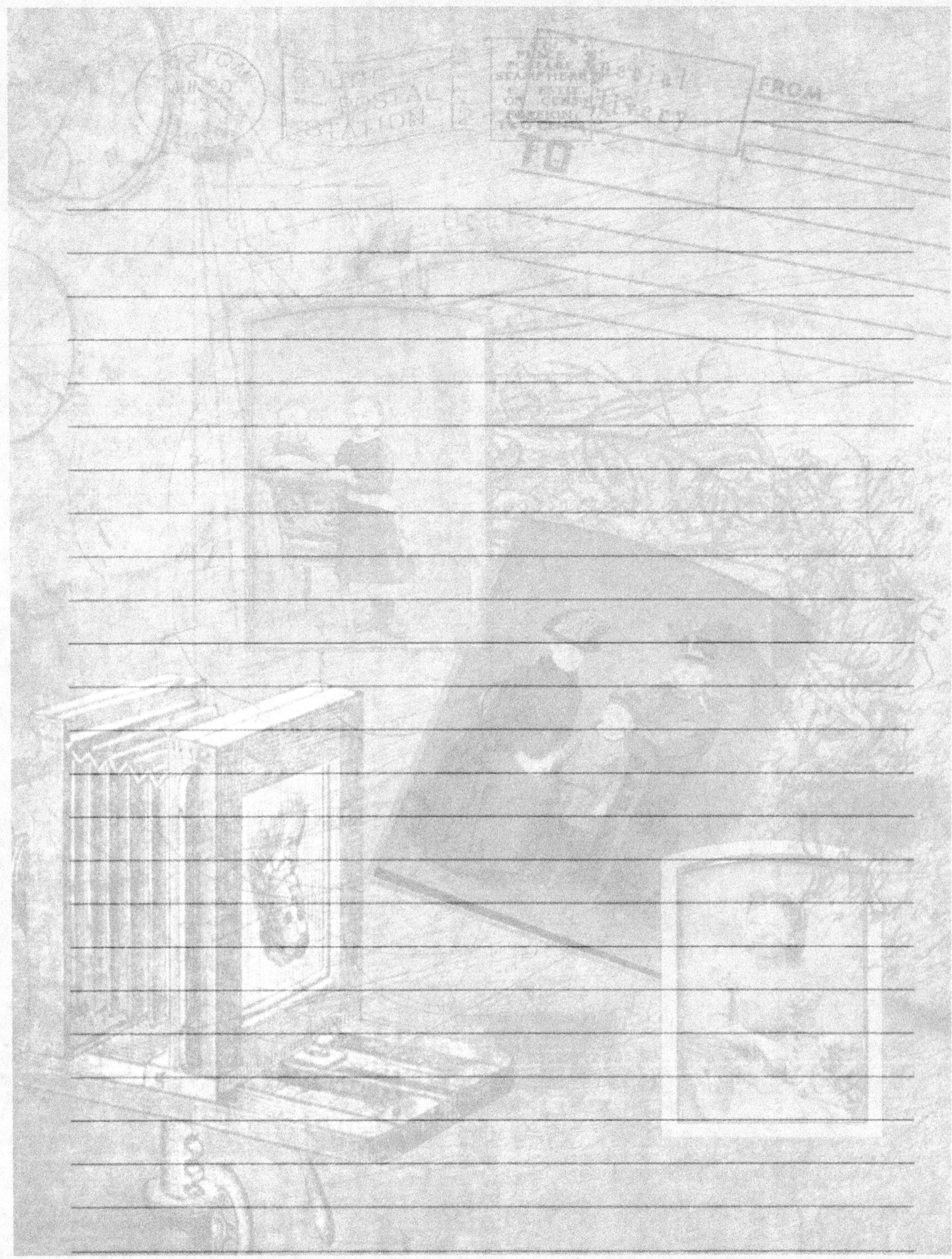

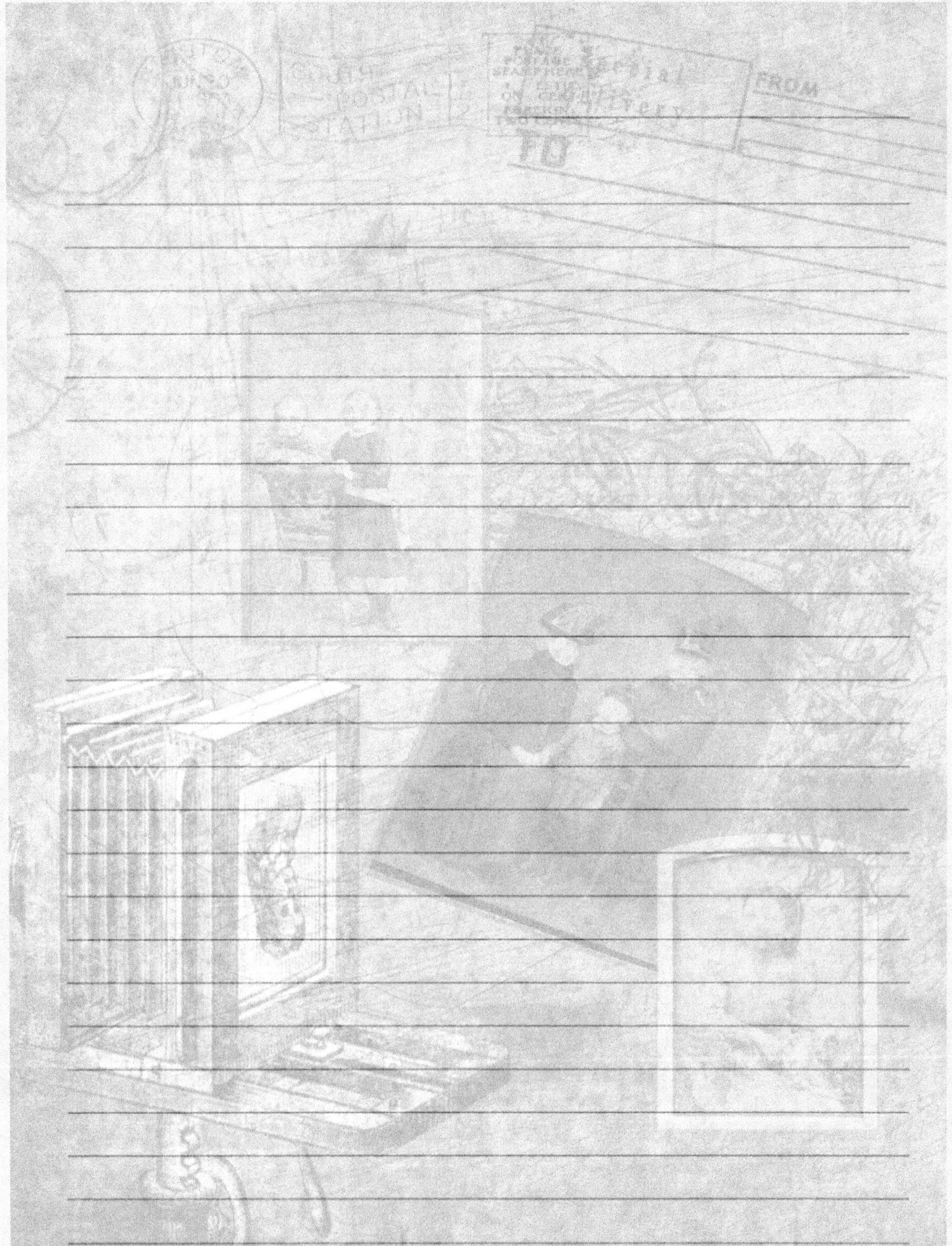

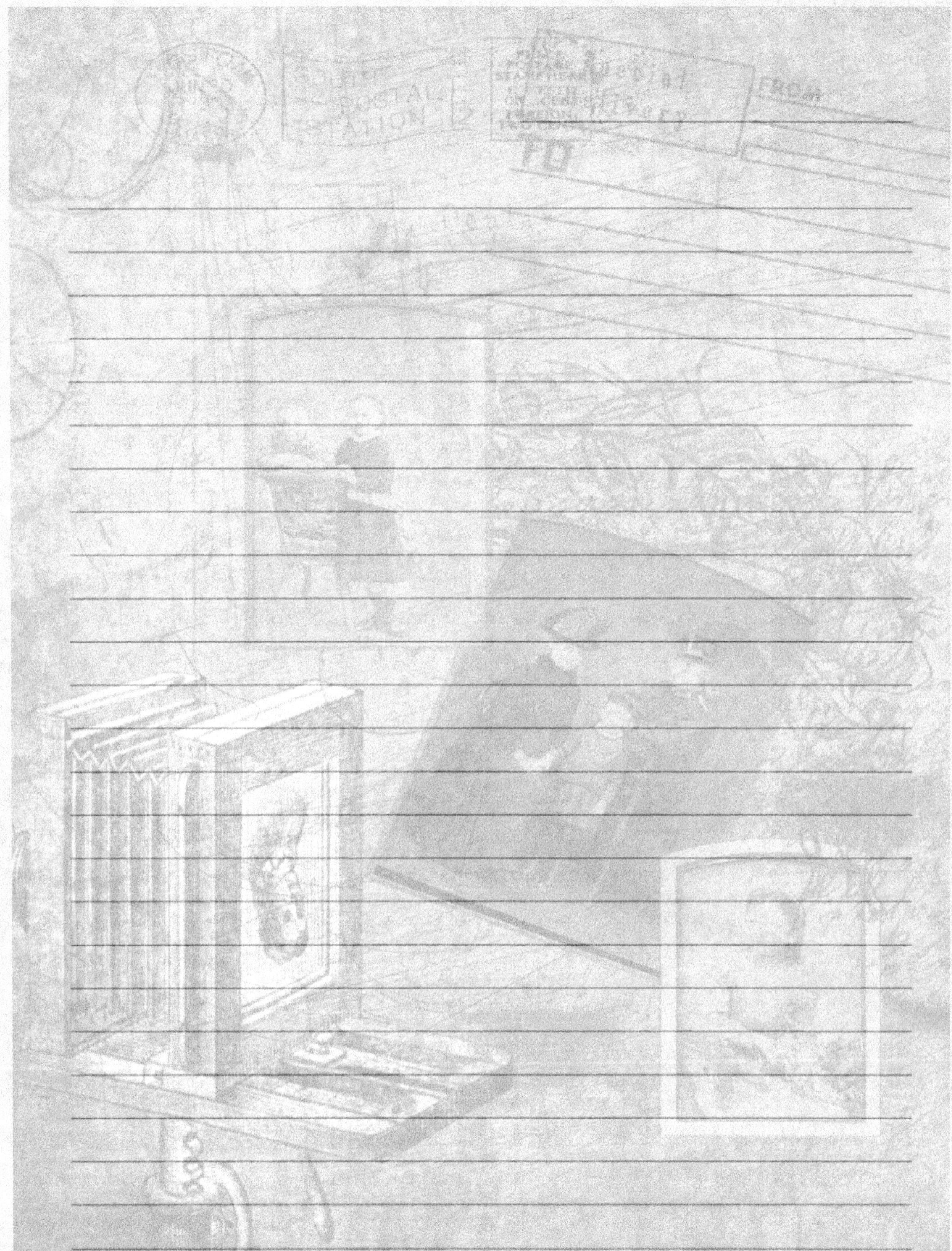

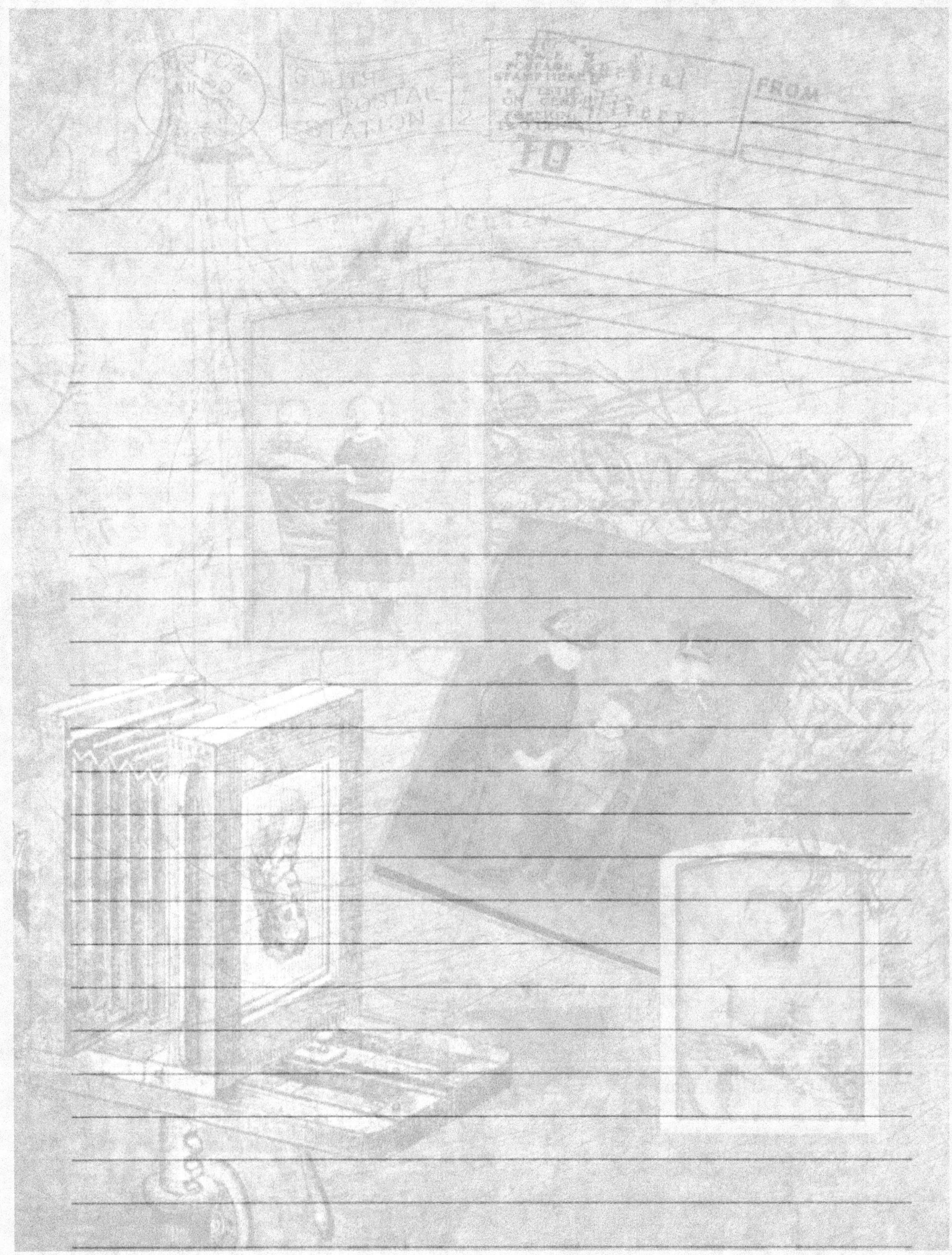

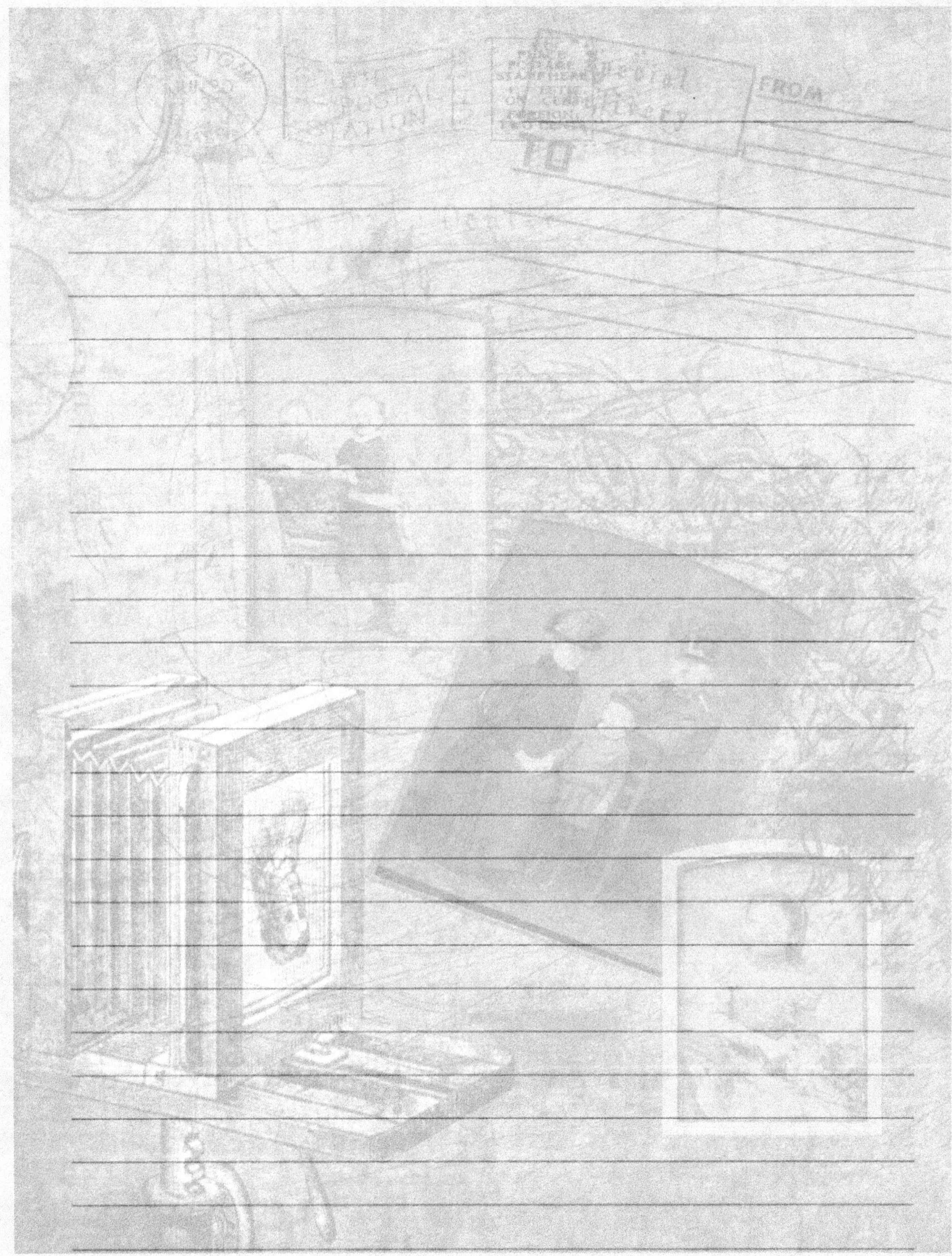

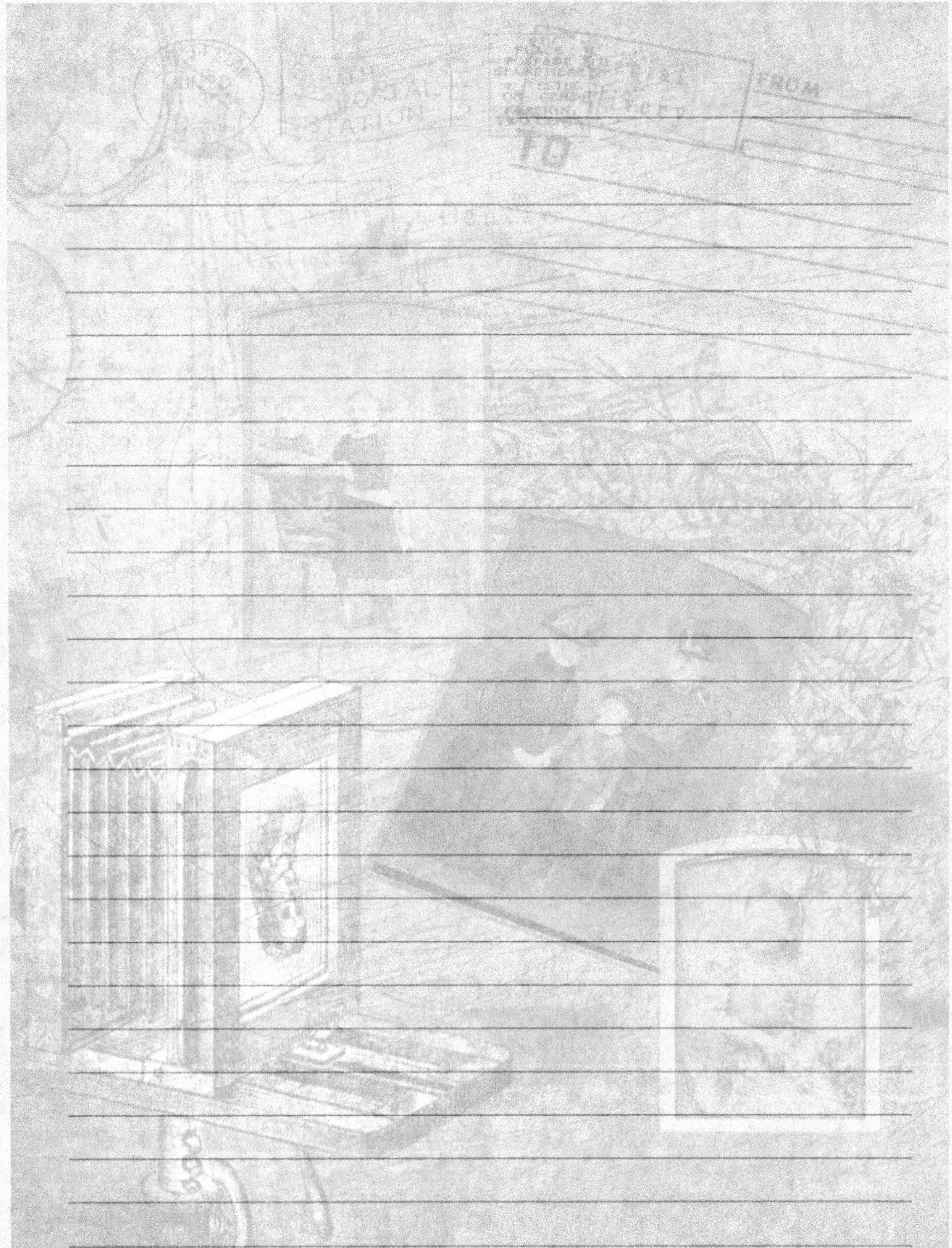

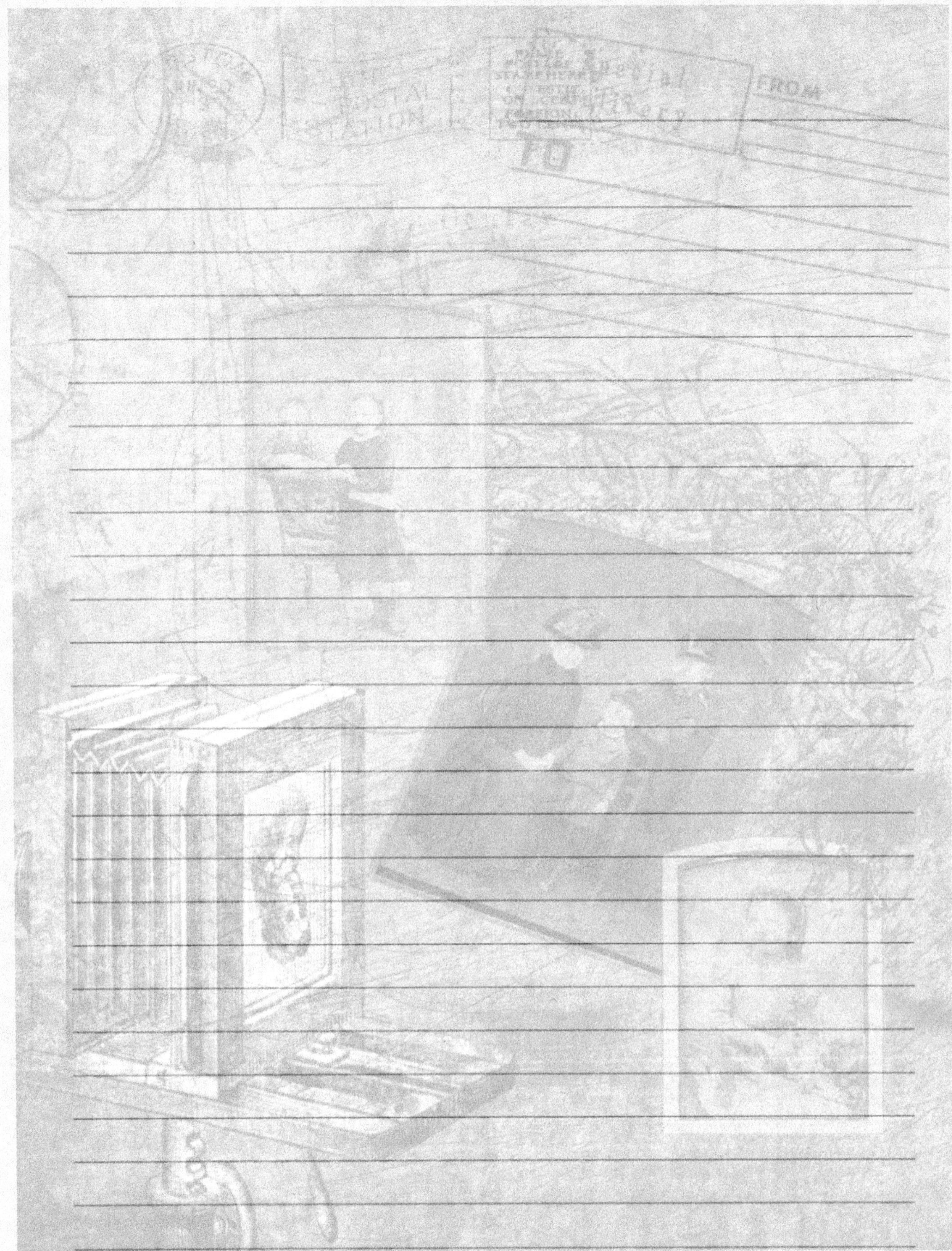

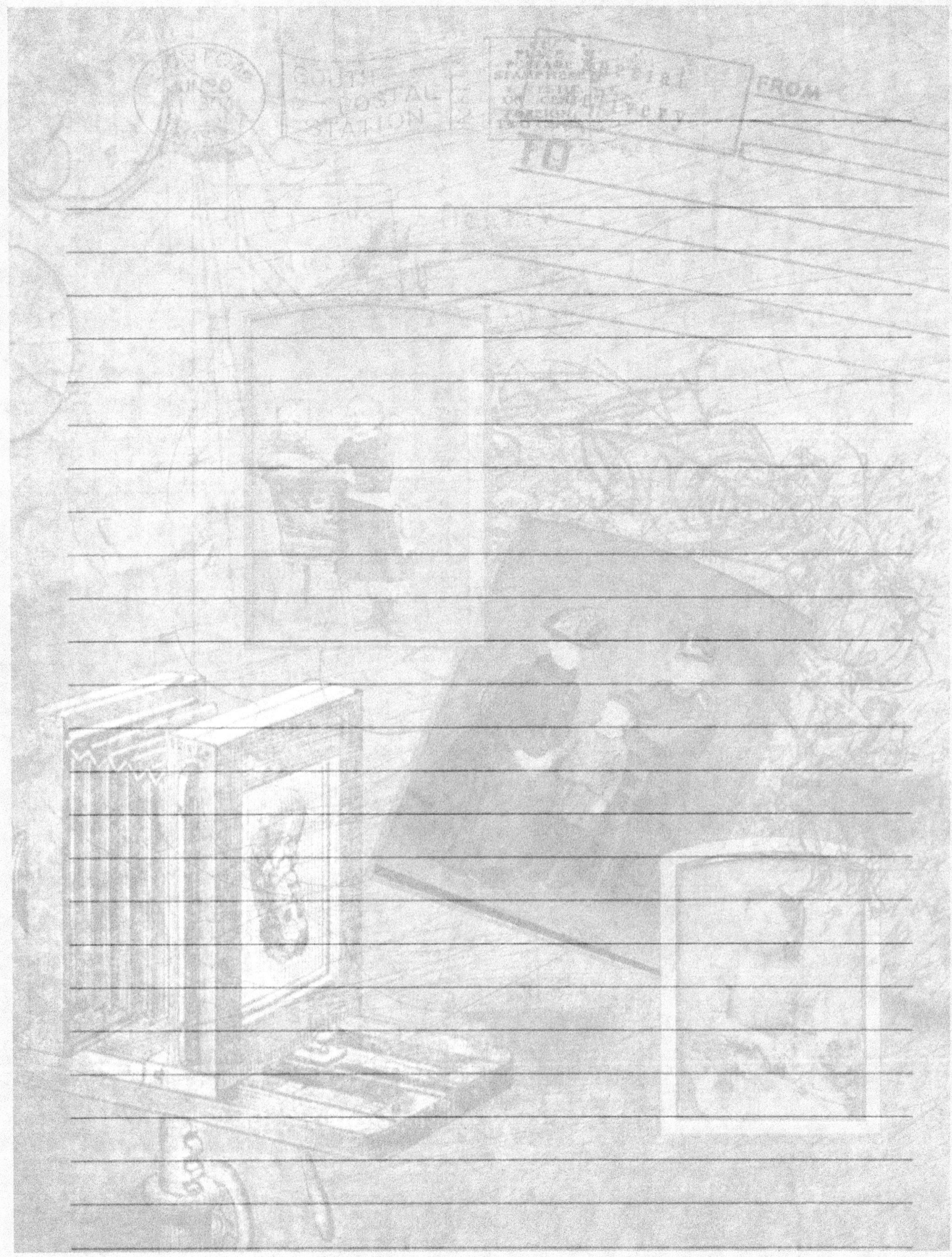

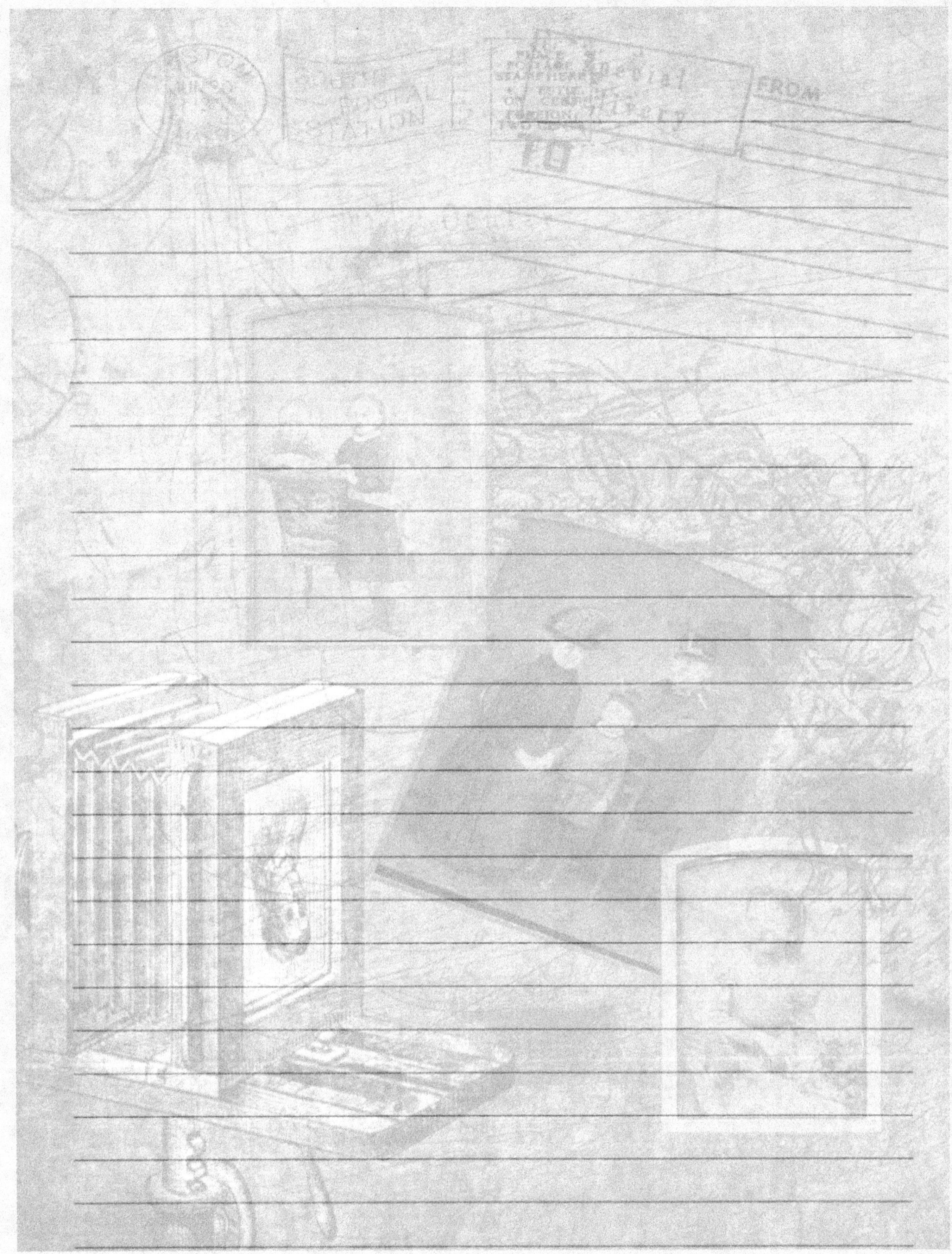

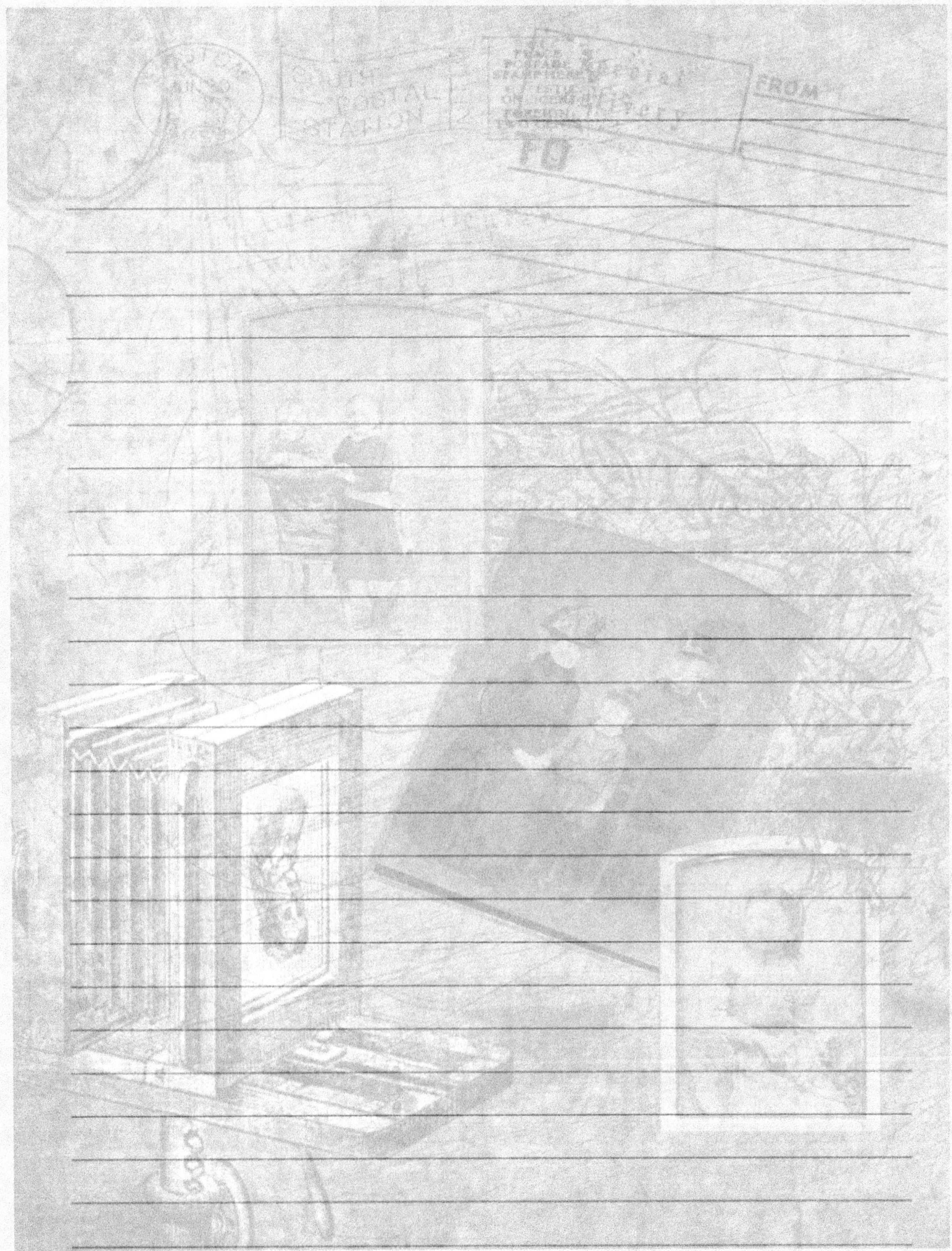

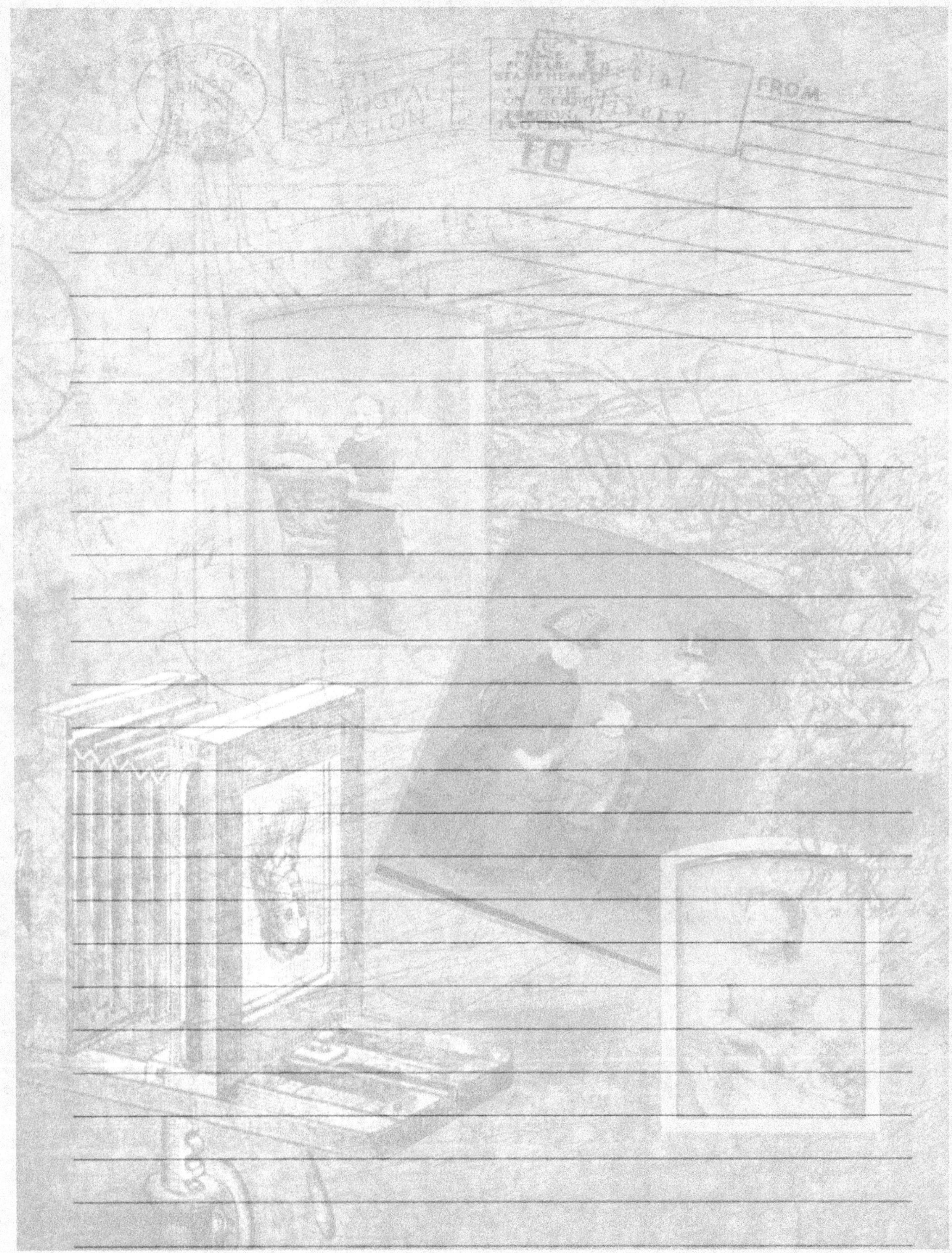

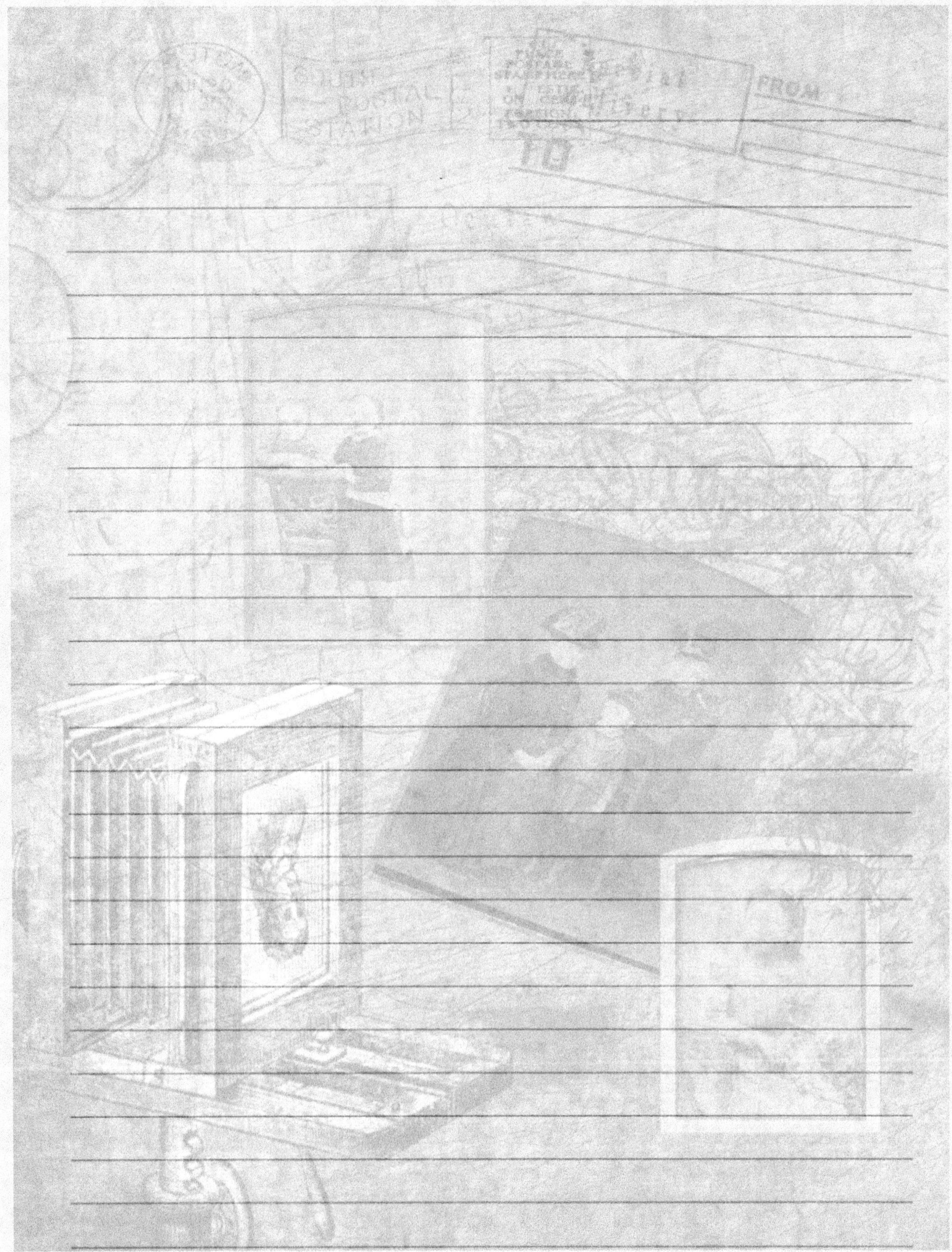

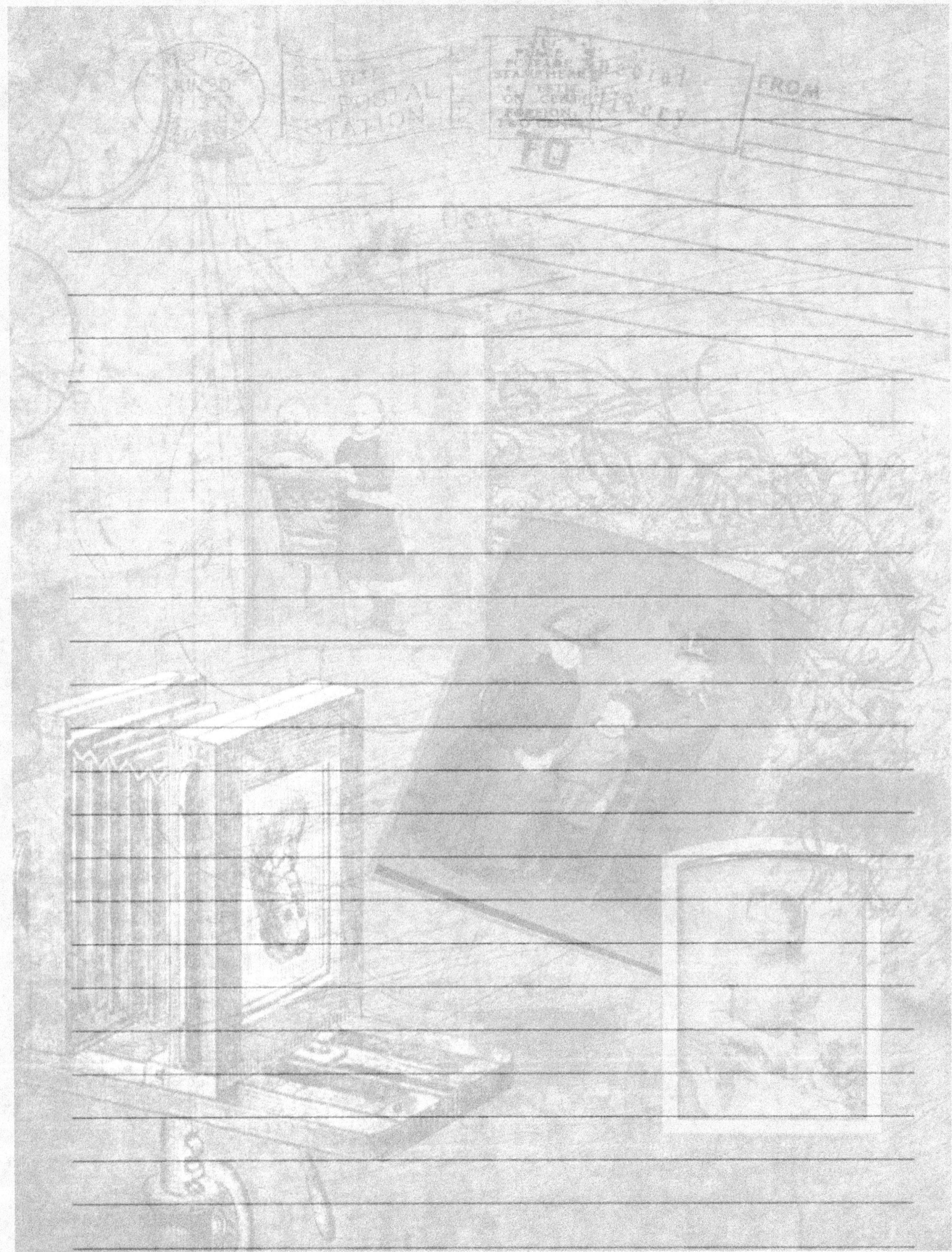

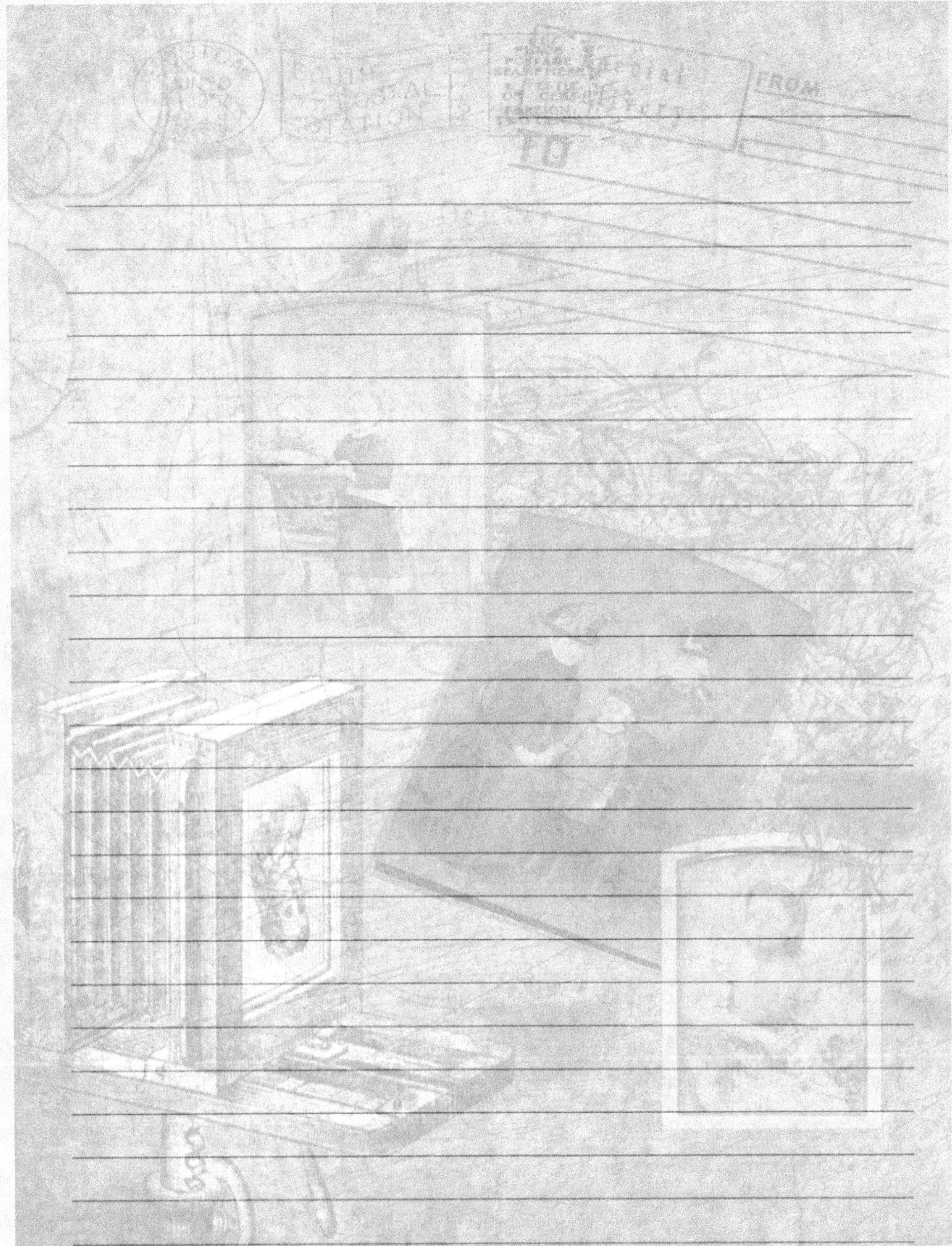